WORLD PUBLIC ORDER OF THE ENVIRONMENT

JAN SCHNEIDER

World public order
of the environment:
Towards an international
ecological law
and organization

UNIVERSITY OF TORONTO PRESS
Toronto and Buffalo

© University of Toronto Press 1979
Toronto Buffalo London
Printed in Canada
Reprinted in 2018

Library of Congress Cataloging in Publication Data

Schneider, Jan.
  World public order of the environment.
  Includes indexes.
  1. Environmental law, International. I. Title.
  K3585.4.S3 1979   341.7'62   78-11712
  ISBN 0-8020-5425-0
  ISBN 978-1-4875-8095-7 (paper)

Parts of this book, in earlier versions, have appeared in the *Yale Law Journal* and *Yale Studies in World Public Order*. Publication of this volume has been made possible by financial support from the University Consortium for World Order Studies and the Yale Concilium on International Area Studies and by a grant to the University of Toronto Press from the Andrew W. Mellon Foundation. The author wishes also to express most sincere gratitude to the many friends who have been so helpful and supportive throughout.

TO MY PARENTS

Contents

ix Contents

Foreword

The broadest reference of 'environment,' in common usage, is to 'all the conditions, circumstances, and influences surrounding and affecting the development of an organism or group of organisms.'* The contemporary concern of the peoples of the world about the quality of their environment appropriately extends, in comparable reach, to the whole complex of resources – atmospheres, waters, and landmasses – which both envelops and forms the material base for all their activities in the shaping and sharing of values. The new international law of the environment, as Jan Schneider's book makes clear, is but our inherited international law about the management of resources (potential values) writ large, with emphasis upon emerging contemporary problems.

It is commonplace among the knowledgeable, requiring only brief allusion by Dr Schneider, that all the resources of the earth-space community are knit together in a maze of intimate ecological interdependences – embracing all such features of the material environment as air, climate, topography, soil, geologic structure, minerals, water resources and access to waters, natural vegetation, and animal life – which condition, and in turn are conditioned by, the institutions and practices by which the individual human being seeks to satisfy social, psychological, and bodily needs and demands. Because of inescapable physical, technological, and utilization unities, the resources of the globe, taken as a whole, are today of necessity as sharable, and as requiring of shared management, as are the resources of a single river valley.

It is scarcely less commonplace, as concerned activists continuously remind us, that highly destructive, and sometimes irreversible, damage is being done to all the resources of our global environment at an accelerating rate. Easy to

* *Webster's New World Dictionary* (2d college ed. 1976) s.v. 'environment'

observe, or to anticipate, are the exhaustion of some vital resources, the pollution of all resources, the shrinkage of open spaces, the spoliation of agricultural lands and spread of deserts, the congestion and deterioration of urban areas, the increasingly rapid extinction of many forms of non-human life, and the destruction of natural beauty. Continuing technological advances multiply the potentialities of destructive impact, and a burgeoning population makes cumulative demands upon increasingly strained resources. The Club of Rome needs to be accurate only in minimal degree for us to know that the quality of life for all is intensely threatened.

It is the principal purpose of Dr Schneider in *World Public Order of the Environment* to create, and briefly to illustrate the application of, a comprehensive conceptual framework which will facilitate clarifying and implementing the common interest of humankind in an appropriate and effective international law of the environment. For her, as the book repeatedly emphasizes, there is an ecological dimension, just as there is a human rights dimension, in every human interaction and in every authoritative decision about such interactions. The standpoint she assumes is that of the observer or decision-maker who identifies, not merely with some single parochial community, but with all individual human beings in all their multiple, concentric, and interpenetrating political communities. From this perspective, she finds it necessary to specify in detail the recurrent, factual problems with which authoritative decision-makers must cope in the global management of resources, to postulate certain overriding general community policies for the guidance of such decision-makers, and to engage in a variety of particular intellectual tasks in the detailed specification of such policies.

The public order problems with which Dr Schneider works include those in relation to the allocation of resources, the regulation of use (including both injurious and productive or harmonious uses), the planning and development of resource potentials, and the control of the access of individuals to resources. The authoritative decision problems she categorizes cover the whole of the making and application of general community policy, including the intelligence, promoting, prescribing, invoking, applying, terminating, and appraising functions. She realistically observes that effective power maintains on the global level a comprehensive constitutive process of authoritative decision entirely comparable to that maintained within national communities.

The overriding policy Dr Schneider recommends for postulation by authoritative decision-makers in coping with particular problems about the environment is not that of some simplistic physical conservation of resources in a pristine, untouched state of nature, but rather that of an appropriately conserving, economic, and constructive employment of resources in the greater production and wider distribution of all values of human dignity. 'The right of all people in present and future generations,' she writes (page 67), 'to "freedom, equality, and adequate conditions of life in an environment that permits a life of dignity and well-being" is, after all, what our concern with the human environment is about.'

The intellectual tasks Dr Schneider recommends for the clarification and implementation of policies about particular environmental problems are several-fold, including:

1/ the specification in detail, from the perspective of an observer identifying with the whole of mankind, of basic general community policies in terms of costs and benefits in relation to all values;

2/ the survey of past experience, including prior trends in decision, at all levels of community, from local to global, in terms of approximation to clarified policies;

3/ inquiry into the factors that have affected past experience and decisions on particular comparable problems;

4/ the projection of developmental constructs about probable future decisions and conditioning factors in relation to particular problems; and

5/ the invention and evaluation of new alternatives in rules, institutions, and decision for the better securing of clarified policies.

It could scarcely be expected that Dr Schneider would be able to bring the framework of inquiry and intellectual procedures she recommends to bear upon a comprehensive review of, and assessment of recent developments about, all problems in international environmental law. The framework of inquiry created in this book, certainly the first of its reach, will, however, be of extraordinary usefulness in the guidance of future studies, and she makes an illustrative and creative application of that framework to a variety of important contemporary problems, offering specific suggestions about desirable future directions in decision. To her work as a whole* she brings a unique 'insider's' experience in inter-state, inter-institutional, and intra-institutional environmental diplomacy, a broad understanding of the relevant social and physical sciences, an urbane and cosmopolitan wisdom, and a deep sense of commitment. The book she has produced could serve as a basis for communication for concerned individuals of very different backgrounds and skills, including lawyers, scientists, economists, businessmen, bureaucrats, diplomats, and politicians, and will greatly benefit the common interests of both professionals and laymen.

---

* Dr Schneider has participated in both the United Nations Human Environment and Law of the Sea Conferences. She is a practising lawyer and also has a doctorate in political science.

One of Dr Schneider's most important contributions is in her insistence upon the need, beyond specifically designed programs for the amelioration of particular problems, of a comprehensive and continuous intelligence function about the environmental dimensions of decision. Over one hundred and fifty states get together and try to draft a Law of the Sea 'umbrella treaty,' or an Outer Space agreement, or a New Economic Order – and there are critical environmental implications that must be faced. The Scandinavian countries reach a new Environmental Protection Agreement, coastal states agree on an action plan to protect the Mediterranean Sea and its resources, African states and other concerned countries try to combat desertification problems in the Sahel region – and international environmental implications are both obvious and highly important. Even bilateral problems and efforts to deal with such problems may have far-reaching consequences and precedential value – thus, the United States and Canada make a Boundary Waters or Weather Modification agreement or jointly search the Canadian north for radioactive debris from a fallen Soviet satellite. And so on. Only comprehensive and continuous intelligence can illuminate such problems in common interest.

All who are concerned for the quality and productivity of the global environment, and its constituent resources, must be indebted to Dr Schneider for an invaluable comprehensive map of what hitherto has been wilderness, uncharted waters, and dense fog.

J. ALAN BEESLEY, Q.C.
Canadian High Commissioner
to Australia, former Assistant
Secretary of State for External
Affairs and Legal Adviser

MYRES S. McDOUGAL
Sterling Professor of Law, Emeritus,
Yale Law School

Preface

Throughout this study, I try to take a comprehensive perspective and search for an overall conceptual framework. At the same time, it is impossible to ignore the exceedingly pressing nature of certain particular environmental problems or the critical time constraints on various options. A compelling sense of urgency is captured in the parable of the lily pond, a French children's riddle recalled in *The Limits to Growth*. The lily plant doubles in size each day; if allowed to grow unchecked, it will cover the pond in thirty days, choking off all other forms of life in the water. So what happens? 'For a long time the lily plant seems small, and so you decide not to worry about cutting it back until it covers half the pond. On what day will that be? On the twenty-ninth day, of course. You have one day to save your pond.'*

This sense of urgency, it follows, is in no way denied or rejected here. There are certain inexorable resource and time constraints limiting decisions and actions by the world public order today, which together have generated the present 'crisis' of the human environment.

But the focus of inquiry here is different. I try, rather, to examine the will and means for dealing with both immediate and long-term environmental problems. International environmental law and organization are viewed as a composite of multiple and interpenetrating decision-making processes, responding to perceived natural and physical limitations and implementing value priorities within the remaining bounds of choice. From analysis of trends in these areas, there appears to be emerging a new world public order of environment and resources; this new order is, however, still in an embryonic stage, and we have far to go towards the goal that I have labelled an 'international ecological law and organization.'

* D. Meadows, D. Meadows, J. Randers, and W. Behrens, *The Limits to Growth* 29 (1972)

PART ONE

Introduction

1

Ecology and political ecology

Once there was man and only one earth; there is still only one earth.

It is today becoming widely recognized that this planet – or, more expansively, the entire earth-space system – is an ecological unity both in the basic scientific sense and in the interdependencies of the social processes by which mankind uses it. The plants, animals (including *homo sapiens*), and microorganisms that inhabit the planet are united with each other and with their non-living surroundings by a network of complex, interrelated, and interdependent natural and cultural components known as the planetary 'ecosystem,' indivisibly uniting the multitude of subsidiary ecosystems. Within this system, man alone has a dual role: both as natural symbiotic component and as purposive decision-maker choosing among a range of possibilities. Thus, in order to predict or mould future world events, it is essential to investigate the latter as well as the former aspect – to understand political ecology as well as the natural laws that govern the physical environment.

The state of the *milieu* within which political decisions have to be made, it follows, is readily apparent. The more specific ecological unities or interdependencies – physical, engineering, and utilization – of the planetary ecosystem make our whole earth-space environment a single shareable and necessarily shared resource. The emerging scientific perspective is that of 'only one earth,' as set forth by Barbara Ward and Rene Dubos:

There is a profound paradox in the fact that four centuries of intense scientific work, focused on the dissection of the seamless web of existence and resulting in ever more precise but highly specialized knowledge, has led to a new and unexpected vision of the total unity, continuity and interdependence of the entire cosmos.[1]

4 World public order of the environment

The implications of this new vision for political decision-making have been most impressively analysed by Harold and Margaret Sprout, who define the emerging 'ecological perspective' on international politics as follows:

The ecological way of seeing and comprehending envisages international politics as a *system of relationships* among *interdependent, earth-related* communities that share with one another an *increasingly crowded planet* that offers *finite* and *exhaustible* quantities of *basic essentials* of human well-being and existence.[2]

These understandings of the nature of ecology and political ecology provide the starting point for the present study.

It is also becoming widely recognized that this planet is gravely endangered, and that its rapidly accelerating degradation is reaching a point of no return – a cusp at which we may arrive within perhaps fifty years if strong measures are not taken to halt the forces of degradation and reorient many of our basic patterns of resource use. Rising demands for improved knowledge and more appropriate measures for environmental protection and preservation were dramatically illustrated in the environmental benchmark of 'Stockholm '72.' Three separate environmental conferences took place in that city in that year.[3] The Dai Dong Independent Conference (its name being derived from an ancient Chinese precept: 'For a world in which not only a man's family is his family, not only his children are his children, but all the world is his family and all children are his') met from 1 to 6 June. The official United Nations Conference on the Human Environment, convening 1200 delegates from 113 countries, was the focal event, and it met for two weeks from 5 to 17 June. Finally, the UNCHE was complemented and paralleled by an unofficial meeting organized by various citizens' groups. Although the substantive developments at both the official and the unofficial gatherings disappointed many participants and observers, what happened at Stockholm is most significant in indicating that, at both élite and mass levels, perceptions of ecological factors and consequent demands on the world decision-making processes are becoming more comprehensive and holistic.

Unfortunately, however, these demands are as yet neither sufficiently perceptive nor sufficiently all-embracing. It is still not widely understood that there are ecological dimensions, just as there are human rights dimensions, to all the authoritative decisions taken in all our political communities from local to global and that the rational making of these decisions requires that a comprehensive ecological perspective be brought to bear on all of them. Similarly, there is not yet sufficient realization that beyond the mere infusion of relevant information about ecological unities and common interests into the traditional flow of decisions, positive and dynamic programs for better protection and more advantageous use of the whole earth-space environment – land masses, oceans, atmosphere, and so on – are crucial. The mere collective understanding that things should be done, in other words, does not necessarily ensure that they can or will be done.

The publicists have not been much more unified and coherent than the politicians in their approach to environmental management problems. There are wide-ranging extremes: from the doomsday politics people, on the one hand, who decry the inescapable predicament of mankind, to the world federalist cheermongers, on the other, who are happily convinced that the so-called 'environmental crisis' is essentially a practical problem of the same order and as susceptible to ordinary techniques of negotiation and functionalist recipes as a host of earlier international difficulties.[4] If the most extreme views of the global environmental situation were fully proved or accepted, there would be no reason for the present work; it would either be irrelevant given the fact of impending and unavoidable disaster or else an unnecessarily arduous and tedious exercise that could have been avoided by letting human nature take its course and world politics follow suit. But, in the face of admittedly incomplete but nevertheless convincing scientific evidence, it is better to heed the warning and assume that there is a real and imminent crisis of global proportions for which immediate and long-term preventive and ameliorative measures must be sought. In this sense, the well-known study of *The Limits to Growth*[5] and several other of the more pessimistic assessments of the situation have been invaluable in issuing a warning and arousing genuine concern.[6] After all, we must take into account not only the probability of being right, but also the possible consequences of being wrong.

This book attempts a broad contextual analysis of the problem. It looks at the ecological or environmental dimension to the 'world public order' or 'international law and organization' – defined here as those features of the world social or political process, including both goal values and implementing institutions, which are protected by law ('law' being used in its broad sense of the expectations and practices of a social or political community). After the dimensions of this public order, a composite of multiple and interpenetrating communities from local to earth-space in scale, have been introduced, the main body of the work presents and evaluates evidence of progress to date in managing the total environment and its various parts. Illustrating this overall analysis are two problem studies, the first expanding upon the regime of state responsibility for environmental protection and preservation, and the second illustrating the organizational problems of resolving environmental disputes.

The evidence to date does show some progress in the evolution in early stages of international environmental law and international environmental organization. Crisis is far from having been averted, however, and it is the purpose of this book to suggest ways of taking greater account of ecology and political ecology, of moving towards an international *ecological* law and organization (the latter term being intended to highlight the interrelationships and interdependencies of resource management).

There is no intent here to marshall evidence either for or against world government in the popular sense of an established, centralized system of administration and policing. I do contend that the present pluralistic world community and public order is beginning to recognize and take account of the ecological dimension to political choices and that this international system can, without centralized coercion, be rendered more ecologically responsive in the future. The argument implies no great love for sovereign states *per se*; they could be superseded by regional or other larger formal agglomerations as the primary focus of power in world politics without basically altering my thesis. Whether or not any degree of federation (or conquest) up to a totally centralized world government is either likely or desirable depends on many military, cultural, and other considerations beyond the scope of this study. How such a government might cope with environmental problems is also beyond the realm of speculation here. My inquiry is slightly more modest. Given the contemporary world public order, how may we preserve and protect the human environment in the calculable future?

2

Environmental public order

The protection and preservation of the earth-space environment is essentially
a public order problem, in the sense that it affects the whole global community
and its multiple and interpenetrating component communities. There are, of
course, innumerable private consequences of environmental policies, but the
nature of environmental protection and management measures remains inescapably an issue of public choice, which may require both collective and individual action. The international community has developed an overall world
public order or process of authoritative decision for making and implementing
choices. The fundamental features of this decision-making process mold and
determine the particular decisions comprising the international environmental
dimension, while at the same time they are themselves being determined by
the latter. Thus, it is necessary to observe the basic parameters of the total
international system in order to understand trends in environmental public
order. Can we identify the major actors possessing authority and/or control in
international environmental decision-making? What is the nature of the interests held by those actors? Is there a logic according to which they act or
fail to act to achieve their common interests, and does it suggest ways of
reorienting the system to take better account of environmental costs and
benefits?

Actors

Although the states-as-sole-actors approach to international politics has long
been discredited by practitioners as well as theoreticians of diplomacy, and
although the idea of states as sole 'persons' at international law has also been
thoroughly undermined,[1] the primacy of the state in contemporary international law and politics remains unchallenged. It has already been remarked

that, in the present interstate system, national governments retain principal authority and control both over the causes and over the ordering of environmental interactions. In consequence, the United Nations Declaration on the Human Environment not only acknowledged that states have the 'sovereign right to exploit their own resources pursuant to their own environmental policies,' but also proclaimed that they have the concomitant 'responsibility to ensure that activities within their jurisdiction or control do not cause damage to the environment.'[2]

Yet there are other highly significant actors. After the United Nations Conference on the Human Environment, the UN General Assembly created a body, the United Nations Environment Programme, to provide general policy guidance for the direction and co-ordination of environmental efforts within the UN system; UNEP was composed of a Governing Council, an Environment Secretariat headed by its Executive Director, an Environment Fund and an Environment Co-ordination Board.[3] In the sectoral fields, virtually every one of the UN Specialized Agencies has ongoing environmental activities, as do the UN Regional Economic Commissions, the World Bank, and many other members of the UN family.[4] In addition, the non-governmental organizations, private groups, and individuals with avowed environmental interests and projects are legion. They range from bodies of world-wide renown such as the International Council of Scientific Unions and the International Union for the Conservation of Nature, through many national and subnational resource conservation and other groups, to individual concerned citizens.[5]

Furthermore, of course, certain transnational associations and other non-sovereign actors – like nation-states – have some effective power over the causes of environmental problems as well as some impact on international decision-making processes. Huge multinational corporations and shipping concerns have command over major portions of the earth's resources and are responsible for large amounts of pollution; their legal status is still fraught with many ambiguities, and their vast wealth is considerably fungible in terms of national and international political power.[6] Even the most genuinely well-intentioned bodies may promote particularly disastrous environmental results; *Humanae vitae*,[7] for example, is far from conducive to wise population policies.

Although various specific actors will be mentioned later in connection with their particular policies or the functional tasks they perform, the job of overall categorization is left for environmentally minded computers. Rather than attempt to provide a directory or 'grocery list' of relevant international bodies, the intent here is simply to re-emphasize the fact that there are multiple actors and political processes of varying territorial compass that have a constitutive effect on each other. International environmental law and organization, like other aspects of world politics, thus represents a composite of many subsidiary processes (deliberate and unconscious) of collective decision-making and practices. The sovereign state is the primary actor in the international arena, but its perceptions and policies are developed in conjunction with many other actors, both international and domestic, and in response to a host of political, economic, technical, and other factors at all levels of interaction.

Interests

Within the world political arena, states and other actors pursue several different types of interests. Endless arguments can be – and, unfortunately, often are – waged among lawyers, political scientists, economists, and others as to the definition and implications of 'public goods,' 'community interests,' 'collective interests,' 'common heritage of mankind,' and similar theoretical constructs. Without aspiring to etymological definitiveness, we can use certain basic definitions and concepts to characterize the relevant interests for present purposes. Various legal and organizational outcomes can then be described in terms of the interests served.

The first basic distinction is between *common interests* and *special interests*. Common interests, also known as collective, group, or sometimes community interests, refer to group demands for values whose achievement is affected by conditions of interdependence and interdetermination. Economists tell us that the satisfaction of a common interest means that what in their parlance is called a public good or collective good has been provided for the group.[8] Special or exceptional interests are, by contrast, those which are destructive of common interests in that they cannot be shared even in equivalences; in other words, their achievement violates conditions of interdependence, imposing unnecessary harm on others.[9] Thus, the basic purpose of any group association, including international organization, is the furtherance of common and the rejection of special interests. All nations and all peoples, to use the broadest example here, have a common interest in the protection and enhancement of the human environment; certain activities, such as excessive pollution of the oceans or atmospheric nuclear testing, clearly serve only special ends and are destructive of the common good or heritage.

Within the overall realm of common interests, actors have both *inclusive interests* and *exclusive interests*. Inclusive or shared interests refer to those that have significant transnational effects, that is, that have an important effect on more than one territorial community. Exclusive or non-shared interests, on the other hand, predominantly affect only one territorial community. In jurisprudential terms, the distinction has been presented by Myres McDougal, Harold Lasswell, and Ivan Vlasic as follows:

Inclusive public order interests:
Demands for values plus supporting expectations about conditions of achievement, the expectations involving high degrees of collective impact upon the relationships referred to by the goals of the world community. Community-wide participation in decision, or a lesser degree of participation by more than one component community of the world arena.

Exclusive public order interests:
Demands for values plus supporting expectations, the expectations involving high degrees of particular impact, compatible with the goal values of the world community, and unaccompanied by high levels of collective impact.[10]

In terms of interest satisfaction this means that with inclusive interests the value automatically expands when the group expands; in the case of exclusive interests, on the other hand, the amount of value to be derived from their satisfaction is fixed in supply and there is therefore competition among group members for exclusive or particular portions of the benefit. Taking note of this property, Mancur Olson, an economist, also made the distinction between 'inclusive collective goods' and 'exclusive collective goods' – the former being shared or non-competitive and the latter market or competitive in nature – in his theory of groups and organizations and their action to achieve common interests.[11] He has since reiterated and enlarged upon his basic argument that not all collective goods entail group indivisibility or 'nonexcludability' and has shown the relevance of this observation to the evaluation of collective performance.[12]

This method of interest differentiation could probably do with some elucidation. States and other actors have both negative and positive common environmental interests in the minimization of environmental injury and in securing constructive use of the environment for the benefit of present and future generations. If these interests are achieved for one, they will be achieved for all, although different actors may benefit in different amounts.[13] As far as the subclassifications are concerned, in the international environmental dimension, the inclusive common interest that comes most immediately to mind is the prevention or reduction of pollution of the oceans.[14] Pollution control has significant transnational effects, and there is no competition in enjoyment of the less polluted waters *per se*. An obvious example of an exclusive common interest is the conservation of many types of fishery resources.[15] Most currently utilized ocean fisheries are now scarce economic resources, and while all users of a common stock have a common interest in its conservation, each one seeks as large as possible an individual share of the total supply of fish and therefore has an incentive to try to reduce the size of the participating group.

It would not do, however, to overwork any of these distinctions. In summary, within the public order, actors possess certain common interests, whose achievement is affected by conditions of interdependency; these interests may be inclusive or shared by two or more actors, or else they may be exclusive in involving high degrees of particular impact. There are very few (if any) 'pure' cases.[16] Moreover, the same interests or goods may appear to be public and common or simply private, and inclusive or exclusive, depending on the perspective from which they are viewed. This circumstance, rather than proving discouraging, may afford increased flexibility of means for achieving or satisfying the interests.

The logic of interest achievement

It should be evident as a result of the above discussion that common interests may be achieved either by collective action or by the activity of a single interested party. We must consider, therefore, both when and how groups can be mobilized to fulfil their shared environmental objectives and under what circumstances individual actors may be motivated to take action in the overall common interest. We have a concomitant concern, of course, with means of preventing actors from pursuing their purely special ends to the irremediable disadvantage of others.

It is a truism that having an interest and fulfilling its objective are not the same. Given recognition of common environmental interests and even assuming rationality on the part of nation-states and other actors, it does not necessarily follow that these interests will be achieved. As Mancur Olson demonstrated in *The Logic of Collective Action*:

It is *not* in fact true that the idea that groups will act in their self-interest follows logically from the premise of rational and self-interested behavior. It does *not* follow, because all of the individuals in a group would gain if they achieved their group objective, that they would act to achieve that objective, even if they were all rational and self-interested. Indeed, unless the number of individuals in a group is quite small, or unless there is coercion or some other special device to make individuals act in their common interest, *rational, self-interested individuals will not act to achieve their common or group interests*. In other words, even if all the individuals in a large group are rational and self-interested, and would gain if, as a group, they acted to achieve their common interest or objective, they will still not voluntarily act to achieve that common or group interest.[17]

This intuitive notion is illustrated by the fact that, although they provide undeniably valuable goods and services, and despite the forces of patriotism, political ideology, and common culture, philanthropic contributions are a notably insignificant source of revenue for modern nation-states; taxes, *compulsory* payments by definition, are needed.

Olson did demonstrate that certain small groups can provide themselves with collective goods without relying on coercion or any positive inducement apart from the collective good itself.[18] This is said to be because in some small groups each of the members, or at least one of them, will find that his sole gain from having the collective good exceeds the total cost of providing some amount of it; Olson labels these 'privileged groups.'[19] The greatest likelihood that the collective good will be created exists in small groups characterized by considerable degrees of inequality – that is, group members are of unequal size or have unequal interest in the collective good. But even in the smallest groups, the collective good will ordinarily not be provided on an optimum scale – the suboptimality being due to the fact that once any individual has provided it for himself, others in the group cannot be kept from consuming; there is accordingly in small groups something of a curious tendency of '"exploitation" of the great by the small.'[20] The single most important point about small groups remains, however, that they may very well be able to provide themselves with a collective good simply because of its attraction to individual members. The larger the group, the less it can logically be expected to further its common interests by relying merely on voluntary action by group members solely motivated by the achievement of common interests.[21]

This logic of collective action has been applied to the international system, based upon the observation that sovereign states tend to behave analogously to private individuals in a market economy, seeking to maximize their national interest.[22] In the world public order, with its multiple and interpenetrating processes of community decision-making, the same shadow obviously falls between the existence or even the perception of common interests and their realization through collective action. The United Nations itself, although widely acknowledged at many levels to represent the common interest in international peace and development and to be generally a good thing, has notorious difficulties in raising funds and gaining active support for its programs. On the present topic, as has been pointed out already, although there is the undeniable common interest of all nations in environmental protection, international environmental law and organization are still inadequate and incomplete in scope and inconsistent and fragmentary in substance.

Taking account of environmental costs and benefits

The most obvious conclusion from this is that, consonant with the principle of economy, environmental problems should be dealt with at the lowest possible level of inclusivity. Bilateral or regional solutions should be sought where viable and not in contradiction to broader common interests. Smaller groups have a higher potential for success both because of their 'privileged' character and because the actions of each member of the group are more noticeable to any other members and their constituent parts, so that more social pressure can be brought to bear to encourage desired outcomes. Just as there may be conflict between subgroups of any organization, nevertheless, such actions may at times come into conflict and have to be resolved at more inclusive levels of international organization.[23]

In addition, because man and his environment are inextricably interrelated by ecosystemic ties and there is therefore an ecological dimension to all human affairs, certain environmental problems have to be dealt with by large numbers of states together or by the world community as a whole. Then what strategies are available within the present world public order?

Assuming that the basic goal of the world community is to take more full account of environmental costs and benefits – to minimize both actual injury and loss of potential gains – there is a limited number of options. The concern or 'crisis' of the human environment can be understood as a problem of what economists call external economies or 'externalities,' which refer to broader non-market interdependencies (positive or negative) that arise inadvertently, without anticipation or without adequate information on the part of the actors involved.[24] There are three basic ways to deal with them.[25] First of all, it may be possible to 'internalize' the problem so that a single actor will take account of all the costs and benefits associated with an activity. In public order terms, this would involve recognizing or carving out an exclusive interest within the overall common interest. Secondly, the actor that generates the effect may be able to come to an agreement with the recipient(s) on the proper level of the effect. This assumes not only that the effects themselves are traceable and measurable, but also that there is some state or other actor that can be held responsible for the conduct. Thirdly, the activity may be controlled directly by some governmental body or by delegation of its powers. This alternative requires a high degree of sustained collective action on the part of the international community, but it may be the only available option with certain types of environmental problems (especially those of a broadly inclusive nature). These approaches are, of course, not mutually exclusive and may often be necessarily complementary to each other.

With the third alternative, reconsideration must be accorded to the problem of the illogic of voluntary action by rational, self-interested individuals or nation-states to achieve common interests.[26] It follows that some form of effective enforcement provisions is essential if states are to be expected to live up to their environmental commitments and apply international environmental law. But such compulsion does not necessarily have to come from central coercive institutions, which in fact the international system palpably lacks. The central community can delegate the authority to apply law to entities with actual policing capabilities – namely, nation-states. The extent to which states will voluntarily exercise such effective control, however, will depend primarily on the degree to which they have particular national self-interests in doing so distinct from their common interests with all others in the law in question and in the overall maintenance of the system. Consequently, it becomes a crucial issue for the world public order of the environment that the authority to coerce be delegated to states whose self-interests are compatible with common environmental interests. The public good – environmental protection and preservation – may thus be achieved as a result or by-product of activities undertaken by states in their own national interest.

Trends in environmental public order and recommendations for the future

The first two strategies identified above rely essentially on methods that have been characterized as 'general' (or 'market') deterrence.[27] They do not require a priori collective decision-making about the correct environmental policies in given circumstances or about tolerable amounts of environmental damage *per se*. They do require the expectations of the world community to change to recognize environmental costs and benefits or, in other words, to recognize the environmental dimension to world social processes. To find out how and how much they are being utilized by the world public order, therefore, one looks to developing trends in international law. If the expectations of the world community are developing in ways that would be expected from the above discussion, it should be possible to offer suggestions as to how international environmental law can be rendered more comprehensive and holistic based upon the public order theory.

The third strategy, by contrast, makes use of the alternative method of 'specific' ('collective') deterrence.[28] In other words, to repeat, it requires sustained collective decision-making and implementation in regard to environmental matters on the part of the international community. There must be an adequate information input and a multi-phase constitutive process to specify and clarify international environmental policies and establish the necessary structures and power configurations to enable them to be carried into effect. In order to assess the performance of the world public order to date from this perspective, it is necessary to look at trends in all these phases of the world process of authoritative decision. Based upon analysis of those trends, one should then be able to formulate and evaluate viable alternatives for the future.

It must be acknowledged that the theory of common interests, public goods, and externalities is far from being highly developed and that the basic information necessary for identifying and assessing environmental variables is also lacking in many areas. Consequently, the present study can at best aspire to provide a very general framework and broad indications of fruitful directions for future policy development, leaving much of the specification and extrapolation for a later series of theoretical as well as diplomatic refinements. Furthermore, criteria for choosing among the three strategies for dealing with environmental effects are difficult to delineate. In some cases, of course, the option of 'internalizing' may be precluded, or virtually so, by the prohibitive costs of excluding non-payers.[29] As far as bargaining in regard to specific activities is concerned, if transaction costs are absent, the same production outcome is obtained under any assignment of clearly specified rights between or among the parties: that is, resources are put to their highest values.[30] In the real world, however, transaction costs do exist, and when they are excessive resources may be allocated by market exchange in a way so strikingly at variance with what would be done if social costs and benefits were accurately recognized that people demand some kind of correction; there will have to be either rearrangement of initial rights so as to reduce the transaction costs or some kind of direct government intervention. But yet, it is not demonstrable that the solution to the problem of determining how much of which public goods to produce can best be solved by government regulation, even assuming the international system were readily capable of reaching and implementing such decisions, unless it is possible to take account of all the tertiary or administrative costs as well as primary and secondary costs of the activities themselves.[31] In short, each method has difficulties and undoubtedly also various 'undesirable features' in terms of predispositions or preferences as to the proper role of government. It remains nevertheless of great value to realize that each alternative does have utility under certain conditions and to try to suggest ways of fostering those conditions, albeit perhaps in a somewhat *ad hoc* manner.

Obiter dictum on minimum and optimum order

From the fact that common interests are realized and public goods created at a sub-optimum level, there follows a clear implication that some minimum acceptable level of environmental protection is all that can reasonably be expected from the present world order, even assuming the exertion of some degree of authorized coercion. This is indeed true, not only because of imperfect ecological and political ecological intelligence, but also because of the inherent nature of the collective decision-making system itself. It is nevertheless worth remarking in closing this introductory section that the common interests, both inclusive and exclusive, of the world community in the use and enjoyment of the earth-space environment relate to both *minimum order* and *optimum order*. By the former is meant the conduct of activities by the processes of persuasion and agreement, with some tolerable amount of unauthorized coercion and destruction; the latter refers to co-operative activity in the utmost production and distribution of all demanded values in world society.[32]

Throughout the present discussion, the focus remains on law and minimum world order – defined as that which is necessary for human survival under 'adequate conditions of life in an environment that permits a life of dignity and well-being.'[33] This should not be taken to indicate closed-mindedness to more favourable solutions, including Richard Falk's projected world order of 'ecological humanism' with greater human solidarity and co-operation among actors.[34] Rather, it should be seen as a consequence of the fundamental perspective here: how do we start from where we are? This is certainly not the best of all possible worlds, but to check its progressive environmental deterioration would itself be a most formidable achievement.

PART TWO

Trends in Development of Environmental Public Order

3

Concepts of international environmental law

What is international environmental law? Since the ecological perspective is all-encompassing and international environmental law has been identified as an integral part of the overall process of world decision-making, the question is unanswerable in this most eclectic form. One can, however, investigate a number of subsidiary polemics bearing on and together clarifying the ultimate one. It is possible to identify trends over time relating to certain perennial problems in the allocation, use, and distribution of resources and to evaluate contemporary conditioning factors both of environmental variables and of the features of the world decision process that are determining the law in each area. Such analysis establishes the range of possible alternatives for future law-making in the light of the limitations inherent in the international system.

The international legal order can be described in a readily comprehensible manner by reference to changing world expectations concerning competence over resources, regulation of resource use, and access of people to resources. The operational questions for the present inquiry then are:

1 / How has competence over resources traditionally been determined at international law? How has this order been changing over time? And what is the general pattern of environmental, technological, and other factors influencing the future development of this order?

2 / How has the use of environmental resources been regulated over time? Why has this regulatory regime become inadequate in the light of the crisis of the earth-space environment? And how can it be improved to embody better resource planning and management in the future?

3 / What are the social and cultural patterns determining how and to what extent groups and individuals benefit from the resources of the human environment? What is wrong with this regime, and are there ways of reducing the inequities under existing conditions of resource scarcities?

Adopting this tripartite framework, it is possible to analyse trends and conditioning factors in principles of international environmental law, so as later to be able to project and evaluate policy alternatives. The first section of this chapter is grounded in basic concepts of competence derived from property law; the second deals with matters usually considered primarily within the province of tort law but also strongly rooted in the law of contract; and the third draws on concepts of equity and human rights considerations in international relations. The approach, nevertheless, is pragmatic rather than legalistically pristine. Relevant concepts and theories are adopted from any and all fields of law and political science as necessary or useful to the inquiries pursued here.

What we gain from this analysis is an overall description of evolving principles of resource law. Use of the notion of 'resources' (including human resources) permits comprehensive treatment of environmental issues, drawing on scientific, legal, political, economic, and sociological data.[1] In sum, international environmental law is an all-inclusive perspective, but analysis of the principles of international resource law provides a structured framework for defining and appraising its most characteristic precepts and contours, which can then be integrated into a unified whole.

International environmental law is such a new and rapidly developing field that this chapter would be hard to write and keep current in any event. Particular difficulties arise, however, because it is being written contemporaneously with the convening of the Third United Nations Conference on the Law of the Sea (UNCLOS III).[2] Much international law regarding pollution, conservation of living resources, and other environmental aspects of oceans management should be clarified and revised during the course of UNCLOS III. Since these negotiations are still likely to go on for quite some time, and since the progressive development of international environmental law will, it may be hoped, keep going forward in many other arenas as well, I have nevertheless decided to proceed on the basis of already discernible trends and main currents of opinion.

Determination of competence over resources

Competence over resources is the right and responsibility of participation in decision-making, conscious or unconscious, in their regard. It must thus be distinguished from the process of use, which refers to the actual benefit or possession of the underlying values derived from the resources. Both may be inclusive or exclusive in the senses defined in chapter 2. Inclusive competence is usually paired with inclusive use and exclusive competence with exclusive use, but such is not necessarily the case. A familiar example of the division is the practice of licensing, where competence or ownership and use or usufruct are divorced for a specified purpose or period of time; government leasing to private individuals is an illustration of a mixed order where the former is inclusive and the latter exclusive.

Principle 21 of the United Nations Declaration on the Human Environment summarizes the international regime of state competence and use of resources:

States have, in accordance with the Charter of the United Nations and the principles of international law, the sovereign right to exploit their own resources pursuant to their own environmental policies, and the responsibility to ensure that activities within their jurisdiction or control do not cause damage to the environment of other States or of areas beyond the limits of national jurisdiction.[3]

There is a fundamental tension in this system. On the one hand, states have a sovereign right to decide policies for the management, development, allocation, distribution, and other matters of exploitation of their own resources; but, somewhat analogous to the *sic utere* principle in domestic law,[4] they also have the correlative duty to ensure that the resultant activities do not injure the environmental interests of others or common environmental interests.[5] They may not, in other words, be in pursuit of solely special interests.

What this formulation represents is a most significant conjunction of two balanced principles of international resource law – permanent sovereignty over natural resources[6] and the doctrine of state responsibility.[7] In order fully to understand its meaning, however, it is necessary first to be able to identify just what it is that states do or can 'own' and what is held as *res communes* (things common to all).[8]

The international legal order of competence over resources inherited from past generations has shown marked preference for the delineation of boundaries in physical media. Resources subject to exclusive appropriation and policy-making have traditionally included the land masses[9] and their closely proximate waters (in particular, internal waters[10] and the territorial sea[11]), their superjacent airspace,[12] the continental shelf down to certain specified limits,[13] and the genetic, aesthetic, and cultural resources within these areas.[14] The resources remaining under inclusive competence have therefore been the oceans, the airspace above the oceans, and the ocean floor,[15] the void of space and celestial bodies,[16] international rivers,[17] Antarctic areas,[18] and some flow[19] and stock[20] resources.

INCLUSIVE RESOURCES

Claims to limited exertions of exclusive competence have in the past been raised and honoured in connection with some specific interests in or uses of basically inclusive resources. The historic recognition of contiguous zones adjacent to coastal states' waters provides an example. The 1958 Geneva Convention on the Territorial Sea and the Contiguous Zone allows a coastal state to exercise, over a maximum of twelve nautical miles breadth, the control necessary to:

a / Prevent infringement of its customs, fiscal, immigration, or sanitary regulations within its territory or territorial sea;
b / Punish infringement of the above regulations committed within its territory or territorial sea.[21]

In addition, states have from time to time claimed and been allowed competence over various supplementary zones – customs zones, fisheries zones, and security zones in the oceans and air defence zones over the high seas.[22]

These claims are all fundamentally derived from the international law principle of 'impact territoriality,' which establishes the competence of a state with respect to events which occur within or have impacts upon its territorial base.[23] Contiguous and other traditional special zones are unlike territorial seas and other exclusive areas in that states do not possess 'sovereignty' – in the sense of supreme political authority and paramount control – over them. Nevertheless, to repeat, for certain specified purposes they do have some recognized limited exclusive interests and resultant policy-making rights and responsibilities – that is, competence – therein.

In the light of the emergence of resource scarcities and as a reflection of rising intolerance of the limitations inherent in the present international decision-making system, traditionally inclusive resources are increasingly being subjected to claims of expanding exclusive competence. There have been relatively minor extensions of claims to exclusivity with reference to the criterion of geographic factors upon which the original inclusive/exclusive divisions were made; and there have been major extensions in terms of the legal consequences approach that is the foundation for the lawfulness of contiguous zones.[24] The most pronounced changes are taking place in the field of oceans law. The former trend is reflected in the growing number of states that are augmenting the breadth of their territorial seas claims; and the latter is shown, among other things, by the propounding of new claims for 'exclusive economic zones,' for coastal state 'residual competence' in matters of pollution control, and for special fisheries regimes beyond coastal waters.

Geographic factors

At the time of the 1958 Geneva Conference on the Law of the Sea, the International Law Commission found with respect to the breadth of the territorial sea 'that international practice is not uniform.'[25] In answer to arguments by some states that international customary law fixes a three-mile limit, the Commission noted 'that many States have fixed a breadth greater than three miles and, on the other hand, that many States do not recognize such a breadth when that of their own territorial sea is less.'[26] It concluded only – in opposition to the two-hundred-mile claims by certain Latin American countries – 'that international law does not permit an extension of the territorial sea beyond twelve miles.'[27] Neither the 1958 assemblage of nations nor a supplementary Geneva conference held in 1960 was able to settle the matter, and it has been stated that the fundamental defect of the Convention on the Territorial Sea and the Contiguous Zone was this failure to define the breadth of the territorial sea.[28]

Since the Geneva conferences, territorial seas have been creeping outward, with many states claiming more than twelve nautical miles.[29] Broad agreement nevertheless now exists on a limit of twelve miles, providing satisfactory resolution of other related issues, such as navigation and overflight of international straits and competence over coastal resources.[30] A potentially highly significant addition to this regime, however, is being put forth by island states. While the 1958 conference did not accept the archipelagic concept, UNCLOS III has accorded recognition to the special status of 'archipelagic states' and allowed them to draw baselines connecting the outermost points of the outermost islands and drying reefs, subject to precise geographic limitations. Under the present formulation, the sovereignty of an archipelagic state would extend to the waters enclosed by these baselines, but ships and aircraft would have a right of archipelagic sealanes passage (separate and distinct from innocent passage) for the purpose of transversing them.[31] It is not clear, however, whether states that are prepared to accept this new concept as part of an overall 'package deal' in a new LOS convention would recognize it absent the successful conclusion of such a treaty.

It should be recognized that oceanic waters are not the only inclusive resource actually or potentially to be affected by augmenting claims to the geographic boundaries of exclusive competence. A major problem arises in connection with international straits, both through and over which all states now enjoy freedom of transit but which would be engulfed within twelve-mile limits. Before some governments were willing to accept even simple provision for twelve-mile territorial seas, they wanted guarantees of suitable corridors of navigation and overflight.[32] Thus, the legal patterns of competence over different resources, just like ecosystemic conditioning factors, are themselves interdependent and, to a large extent, inseparably so.

Functional and zonal criteria
Geographic factors have, of course, never been the only reason for the traditional international law division of resources between inclusive and exclusive competence. Furthermore, it is undeniable that the factor of propinquity remains crucial to the new functional and/or zonal concepts. The central point to note with regard to the new claims of various types of competence zones, however, is the progressive shift from resource-specific to consequence-specific criteria. Instead of dealing with the oceans and other resources *en bloc* or as total physical media to be carved up geographically, in other words, the trend is increasingly towards treating these resources as bundles of multiple potential uses, some of which are separable for decision-making purposes. The real function of the zonal concept, therefore, can be seen as serving as a safety valve from the rigidities of line-drawing by boundaries of territorial 'sovereignty' (as is, for example, the case with the territorial sea): it permits the satisfaction of particular reasonable demands through exercise of limited authority and control which does not endanger the whole gamut of inclusive community interests. Claims for each of the following three types of regimes represent new perspectives on various aspects of the international 'environmental' problem.

Exclusive economic zones
These coastal zones have indisputably achieved acceptance in the law of the sea negotiations. The concept of an 'exclusive economic zone' or 'patrimonial sea' had many champions over time. It was endorsed by Caribbean countries in the Santo Domingo Declaration of 1972, won the support of many African and Asian developing nations at several regional meetings, and finally came to be accepted by several advanced industrial states.[33] Theoretical antecedents of the idea have been traced back to the Truman Proclamations on the continental shelf and on fisheries of 28 September 1945 and to the 1952 Declaration of Santiago by Chile, Ecuador, and Peru establishing the 'Maritime Zone.'[34] Whatever the past history, and despite the serious problems that are bemoaned by landlocked, shelflocked, and other geographically disadvantaged states, they have clearly been accorded legal recognition, and most coastal states will undoubtedly legislate their own exclusive zones.

The precise legal status of the exclusive economic zone was, however, long one of the leading possible 'conference breakers' in the current law of the sea negotiations. At UNCLOS III, the major maritime nations have steadfastly maintained that the economic zone is a part of the high seas, subject to certain coastal state rights and jurisdictions. Many coastal states have, however, for their part been equally adamant in regarding the zone as *sui generis*, distinct from the high seas. At the sixth session of the conference, in a compromise 'package deal' negotiated by a small group of about fifteen heads of delegation, an accommodation seemed close to having been reached on this basic status issue: on the one hand, the zone is to be a distinct area characterized by a specific legal regime 'under which the rights and jurisdiction of the coastal State and the rights and freedoms of other States are governed by the relevant provisions of the present Convention'; but, on the other, there shall nevertheless be preserved for all states, whether coastal or land-locked, traditional high seas freedoms 'of navigation and overflight and of the laying of submarine cables and pipelines, and other internationally lawful uses of the sea related to these freedoms,' including those associated with the operation of ships and aircraft.[35] The seventh session came very close to consensus on several crucial related issues, including the questions of the rights of landlocked and geographically disadvantaged or other states in the zone, and of third-party settlement of disputes relating to exercise of coastal state sovereign rights in the exclusive economic zone; but final resolution of these problems awaits decision on a formula for determining the outer edge of the continental margin. In short, the broad outlines of a comprehensive, creative new regime have emerged, but many of the functional and political accommodations have yet to be worked out.

Off-shore pollution control

Marine pollution can come via several routes: pollution from land-based sources (directly through watercourses or by way of the atmosphere), pollution from activities concerned with the exploration and exploitation of the sea-bed (both within and outside the limits of national jurisdiction), vessel-source pollution (through intentional dumping and discharge as well as from accidents), and others (for example, nuclear testing). A fundamental issue facing the Third Committee of UNCLOS III (whose mandate includes the negotiation of draft articles on the protection and preservation of the marine environment) has been the question of off-shore reach of national antipollution legislation and enforcement powers. It is not being settled by the traditional approach to law of the sea of looking at the geographic status of the area involved, but rather has been worked out topic by topic through review and balancing of the interests involved.

When the present law of the sea negotiations began, there was very little conventional or customary international law concerning the prevention and control of pollution of the marine environment. The provisions drafted by UN-CLOS III have gone a long way towards filling this legal vacuum. As a foundation, the international community has achieved widespread consensus on certain basic provisions for a new 'umbrella treaty.' Included among them is the general obligation to refrain from polluting the marine environment. States are also in accord not to transfer pollution from one area to another or transform one type of pollution into another, and agreed on the need for certain measures of global and regional co-operation and technical assistance. They will additionally, for the first time in treaty form, assume positive legal responsibility to co-operate in international ocean monitoring programs and to assess the impacts of their activities on the marine environment.[36]

Looking at the specifics of the legal regime for implementation of these principles, the provisions as to international rules and national legislation to prevent, reduce, and control pollution of the marine environment have been drafted according to the various sources of pollution.[37] As to pollution from land-based sources, states are to establish national laws and regulations, taking into account internationally agreed rules, standards, and recommended practices and procedures; in establishing the global and regional norms, they are to take account of 'characteristic regional features, the economic capacity of developing states and their need for economic development.'[38] They are to take similar legislative action with regard to pollution of the marine environment from sea-bed activities subject to their jurisdiction (although without the double standard according to levels of development), and through the new International Sea-Bed Authority and individually to do the same with regard to pollution from exploration and exploitation activities in the international 'Area.' No dumping within the territorial sea or exclusive economic zone or onto the continental shelf can be carried out without the express prior approval of the coastal state, which can regulate and prevent such dumping.

The standard-setting provisions regarding pollution from vessels were especially controversial and are extremely complicated: first, in the territorial sea, coastal states may exercise their sovereignty to establish antipollution laws and regulations, provided that such laws and regulations 'shall not apply to the design, construction, manning or equipment of foreign ships unless they are giving effect to generally accepted international rules or standards';[39] second, in the exclusive economic zone, they can establish certain other laws and regulations giving effect to generally accepted international rules and standards; and third, there are certain supplementary provisions for 'special areas' within the exclusive economic zone with particularly sensitive oceanographical and ecological conditions. Finally, there is also a draft article on pollution from or through the atmosphere.

Turning from standard-setting to enforcement competence, UNCLOS III has developed several new concepts. While traditionally the province primarily of the flag state (state of registry of a vessel), enforcement powers with respect to pollution prevention and control will now lie also with port and coastal states.[40] Flag states will undertake additional obligations to ensure compliance by their vessels with applicable international rules and standards. In addition, port states will be empowered to undertake certain enforcement procedures in respect of discharge violations – even those occurring outside their internal waters, territorial seas, or exclusive economic zones. Also, under certain carefully delineated circumstances, coastal states will be able to cause certain proceedings and other measures to be taken in respect of violations of national laws and regulations or applicable international rules and standards. Enforcement powers are, however, subject to highly detailed safeguards to preserve freedom of navigation, which some environmentalists think are excessive.[41]

Ice-covered areas are special and are dealt with in a separate and distinct article.[42] In this, the states at UNCLOS III have acknowledged the uniqueness of the Canadian Arctic Waters Pollution Prevention Act of 1970, which, while avowedly functional or zonal in its approach, has not been found amenable to ready classification in any more general political or environmental category.[43]

Special fisheries regimes over anadromous species
Fish have demonstrated themselves to be singularly unconcerned with jurisdictional boundaries drawn on maps. Since fishermen go after the fish, and all kinds of ecological, economic, social, legal, and political consequences come in the wake of these activities, the present trend in deciding competence over fisheries resources is to follow the lead of the fish. As a result, under the terms of the 1958 Convention on the Continental Shelf, coastal states exercise exclusive rights over living organisms defined as sedentary species (immobile or in constant physical contact with the seabed or subsoil at the harvestable stage).[44] That leaves open three types of competence question: competence over coastal species (non-sedentary, free-swimming species which inhabit the nutrient-bearing areas adjacent to the coast of one or more countries – that is, most kinds of fish), anadromous species (a special kind of coastal species that breed and begin to grow in rivers, migrate far out to sea, and finally return to their rivers of origin to reproduce and end their life cycles – for example, salmon), and highly migratory oceanic species (which may range far and wide through the high seas – for example, tuna). (Marine mammals, such as whales and porpoises, also pose similar types of juridical questions.)

At UNCLOS III, there is now general consensus that the coastal state can determine and allocate the allowable catch of living resources in the two-hundred-mile zone. As to highly migratory species, on the other hand, it has become obvious that the coastal state and other states whose nationals fish in the region will have to co-operate directly or through appropriate regional or other international organizations to manage and conserve them. As regards anadromous species, however, special problems arise given the nature of their migratory patterns.[45]

Maintaining anadromous species involves difficult and costly economic and other choices, such as investment in artificial spawning channels, providing temperature control of some rivers, and halting hydroelectric development on the migratory routes. Such conservation measures can, of course, only be taken by the state of origin and return, and it must have some incentive to do so. Yet on the high seas, through their ocean years, the anadromous species of different continents meet and comingle, and it is impossible to identify the stocks of any particular country and river. In order to ensure the return of the right number of spawners to the spawning stream, fish such as salmon need to be protected on the open oceans against overfishing. Also, from the point of view of food production, a maximum harvest can only be realized as the fish are close on shore, approaching their home rivers.

For all of these and other reasons, states at UNCLOS III have decided to allow fishing for anadromous species only in waters landwards of the outer limits of the exclusive economic zone (with some specified exceptions). The state of origin is to ensure their conservation and may, after consultations with other interested states, establish total allowable catches for stocks originating in its rivers. Where there is intermingling of stocks, neighbouring states of origin are to co-operate with regard to their conservation and management. Where appropriate, arrangements are to be made through regional organizations.[46]

By banning most fishing for anadromous species outside the economic zone and allowing coastal states to regulate their fishing within, UNCLOS III has effectively created another kind of functional exclusive competence which reaches far out into the high seas. But again, it is to be a carefully circumscribed and limited competence for particular conservation purposes.

Possible countertrends

While the most noticeable trend has been these large-scale incursions of exclusive competence in formerly inclusive areas, some resources or resource uses remain inclusive. This may be because of the physical nature of the resource, because it is economically infeasible to carve out exclusive competences, or because of certain preferences on the part of the world community for wider sharing.

An illustrative example in the first category is the necessarily shared weather and climate of this planet. Recommendation 70 of the UN Conference on the Human Environment indicated that decision-makers were at least aware of the problem in this domain; it recommended that 'governments be mindful of activities in which there is an appreciable risk of effects on climate' and to this end carefully evaluate the likelihood and magnitude of climatic effects and consult fully with other interested states when contemplating or implementing activities carrying such risk.[47]

An important step towards implementation of this recommendation is represented by the conclusion of the Convention on the Prohibition of Military or Any Other Hostile Use of Environmental Modification Techniques, under which each state party undertakes 'not to engage in military or any other hostile use of environmental modification techniques having widespread, long-lasting or severe effects as the means of destruction, damage or injury to any other State party'; moreover, the parties further undertake 'to consult one another and to co-operate in solving any problems which may arise in relation to the objectives of, or in the application of the provisions of, the Convention.'[48] As used in the new treaty, the term *environmental modification techniques* refers not only to means of weather and climate modification but more generally to 'any technique for changing – through deliberate manipulation of natural processes – the dynamics, composition or structure of the earth, including its biota, lithosphere, hydrosphere and atmosphere, or of outer space.'[49]

As to other resources remaining under inclusive competence, in the second category would come the continued preservation of freedom of the seas and airspace above them as regards transportation and communications uses even within two hundred miles offshore. While it would perhaps not be impossible to establish boundaries in the oceans and air, the resulting burdens on the flow of trade and ideas are found too costly in terms of common community interests. And in the third category might perhaps be the attempt to establish an international regime for the deep sea-bed and its minerals. It is conceded to be economically feasible to stake out and maintain geographical claims for such mining activities, but countervailing political factors – usually subsumed under the overall conceptual label of the 'common heritage of mankind' – are militating in favour of preserving its inclusivity. More will be said on these and related developments subsequently, but for the moment the important point is the basic fact that these competence questions are now being approached on a functional basis in terms of the legal consequences of their assertion, rather than along simplistic geographic or geophysical lines.

EXCLUSIVE RESOURCES

The traditional legal order in regard to exclusive resources has been essentially a *laissez-faire* system oriented towards the unchecked competence of nation-states. The right of a state to control the exploitation of resources within its territory is one of the basic components of state sovereignty. This right received explicit recognition by the UN General Assembly as far back as 1952, when it passed a resolution pointing out that 'the right of peoples freely to use and exploit their natural wealth and resources is inherent in their their sovereignty and is in accordance with the Purposes and Principles of the Charter of the United Nations.'[50] Since then, there have been several reaffirmations by the General Assembly and other international bodies.[51] Principle 21 of the Stockholm Declaration not only reiterated this generally accepted notion, but also gave the blessing of the Charter and the principles of international law to the right of a state to exploit these resources 'pursuant to its own environmental policies.'[52]

The potential for the disruption of the world economy and society resulting from unchecked sovereign prerogatives was dramatically illustrated, however, in the world 'energy crisis' of 1973–4. There would have been a real and pressing resource-scarcity problem in view of the high consumption rates in several industrial countries even without the Arab embargo, and there is a common interest among all nations in conserving scarce energy reserves. But the economic and other chaos that resulted from the sudden cut-offs might well have been averted or at least greatly mitigated by far-sighted planning with adequate time for implementation. In order to avoid such disastrous consequences in the future, oil consuming countries have attempted to take common effective measures to meet emergencies through their Agreement on an International Energy Programme,[53] as a supplement to various national drives for energy self-sufficiency and other measures of adjustment. The Organization of Petroleum Exporting Countries has since tried to press the advantage of OPEC members by raising oil prices still further.[54]

Also, of course, individual governments may take actions motivated by environmental incentives that have profound disadvantages for other states. Some countries, just as an example, have been wary about allowing supersonic aircraft to land on their territory, worried about the many possible adverse effects of such transportation, while other countries have made very substantial investments in the development of supersonic transport.[55]

Yet despite the strength of these sovereign prerogatives, even with respect to the land masses, a regime of modestly inclusive competence has been developing to serve the interests of a world economy and society. The most important prescriptions in this regime are those popularly described as 'private international law' (which confer jurisdiction on the courts of a state to protect its nationals and other interests from injury, including injury effectuated beyond its boundaries)[56] and as 'responsibility of states' (which impose certain limitations on the competence of states over aliens with respect to events otherwise within their jurisdiction).[57] Principle 21 makes clear that the latter concept must now be understood to include state responsibility for *environmental* damage, in order to reconcile national resource and environmental interests with those of the international community collectively.

There are also areas where old notions of exclusive competence are being expanded in conjunction with expansion of inclusive competences. Weather and climate modification is probably one of these. States have exclusive competence over the air and atmosphere superjacent to their land masses and territorial waters, and for that reason and based upon traditional concepts of impact territoriality they must almost necessarily be understood to have a vested right and interest in preservation of their own weather and climate. This has been implicitly recognized, for example, in the 1975 Canada–United States Agreement on the Exchange of Information on Weather Modification Activities.[58] An obvious deduction from it is that actors must pay for the harm they do to the weather and related interests of others, but some weather modification efforts may benefit some states while adversely affecting states not in the scheme. The problem then becomes legally more and more complex, and its inclusive and exclusive aspects are progressively merged.

An increasingly functional approach to resource law-making is not itself inherently jurisdictionally directional (that is, necessarily dictating either inclusive or exclusive competence). Moreover, the 'inclusive' and 'exclusive' denominations themselves must be seen as a continuum of interests and competences rather than discrete categories. Abandonment of the old sectoral boundaries in favour of new functional criteria is largely inevitable in light of modern technological developments and their effects on ecological interdependencies. The inclusive or exclusive character of many new particular competences and the degree thereof are still largely left open to political choice; some theoretical and practical considerations that may influence such choice are examined in chapter 5.

Regulation of use of resources

The general international community has a common interest in minimizing losses, inadvertent or deliberate, that inevitably attend transnational interactions in resource use. It seeks further to secure potential gains through facilitating the productive and harmonious use of resources. Finally, planning and development of new uses can be undertaken to enlarge the scale of potential benefits.

There is no automaticity in the international law-making process for regulating the use of resources. In fact, efforts to formulate regulations from the point of view of common global environmental interests not only must face the inertia of large-group dynamics but also must often overcome the opposition of powerful vested interests. The world public order is nevertheless demonstrating some movement towards increasing measures for environmental protection and preservation.

CONTROLLING INJURIOUS USE

Controlling injurious use of resources is primarily a matter of reducing or taking better account of negative externalities or external diseconomies. Among other things, however, various equitable considerations to be discussed below (see 'Access of people to resources,' 66–74) do inject certain additional goals or at least restraints into the balance. Basically, nevertheless, regulation towards this end aims at reflecting and controlling all kinds of undesirable environmental effects of ongoing activities.

Resources inclusively enjoyed

Long before the international upsurge of concern with 'the environment,' states began to worry about pollution – or, at least, about certain types of pollution. They also concerned themselves with the matter of establishing liability and fixing compensation for any damage that might occur to their own national resources as a result of such pollution. This trend has continued and broadened. In recent years, not only have states sought to control many more types of pollution, but they have also exhibited increased solicitude for the comprehensive protection of common resources *per se* as well as their own exclusively territorial interests. In addition, there has been some initial movement away from the principle of 'floating sovereignty' or 'roving sovereignty' (whereby only flag states have jurisdiction outside the territorial seas of others and, for some activities, even within them), and comparable developments in other domains. The sequence and timing of developments varies, but emerging overall patterns or trends can be seen in the law of the sea, space law, the law of Antarctica, and that of international rivers.

Oceans

The current legal thrust for controlling injurious use of the oceans probably began with the 1954 Convention for the Prevention of Pollution of the Seas by

Oil.[59] The basic rule of this convention is that discharges containing more than one hundred parts per million of 'persistant oils' (a phrase that includes crude and heavy fuel oils but excludes the highly toxic refined products known as 'light oils') must occur outside 'prohibited zones' – generally speaking, the ocean area lying within fifty miles of the nearest coast. Prosecution for offences is left to the exclusive discretion of flag states, and no provision is made for compensation for damages suffered. There are obvious defects to this sort of arrangement, but at least it was a start.

This initial measure was soon bolstered by some provisions of the 1958 Geneva Convention on the High Seas, including:

Article 24:
Every State shall draw up regulations to prevent pollution of the seas by the discharge of oil from ships or pipelines or resulting from the exploitation and exploration of the seabed and its subsoil, taking account of existing treaty provisions on the subject.

Article 25:
1 / Every State shall take measures to prevent pollution of the seas from the dumping of radio-active waste, taking into account any standards and regulations which may be formulated by the competent international organizations.
2 / All States shall co-operate with the competent international organizations in taking measures for the prevention of pollution of the seas or air space above, resulting from any activities with radio-active materials or other harmful agents.[60]

Around the same time, related issues not directly concerned with state responsibility were the subject of international negotiations. The 1957 International Convention Relating to the Limitation of the Liability of Owners of Seagoing Ships[61] and the 1962 Convention on the Liability of Operators of Nuclear Ships[62] provide for a regime of strict or 'absolute' liability for such shipowners and operators but limit that liability to $207 per ton in the former and approximately $100 million in the latter instance. These conventions are not, however, specifically designed to deal with pollution damage to the oceans in and of themselves.

States did perceive that this legal order was sorely incomplete, and the Intergovernmental Maritime Consultative Organization (IMCO) called an International Legal Conference on Marine Pollution Damage. This conference, which met in Brussels from 10 to 29 November 1969, again concentrated on oil pollution. It prepared and opened for signature two important conventions: the International Convention Relating to Intervention on the High Seas in Cases of Oil Pollution Casualties (the IMCO 'Public Law' Convention)[63] and the

International Convention on Civil Liability for Oil Pollution Damage (the CLC or 'Private Law' Convention),[64] which at last came into force on 6 May and 19 June 1975 respectively. The conference also passed a Resolution on Establishment of an International Compensation Fund for Oil Pollution Damage,[65] accompanied by a Resolution on Report of the Working Group on the Fund.[66] Finally, some consideration was given to broader problems, and a third resolution was passed, on International Co-operation concerning Pollutants Other than Oil[67]; it merely recommended, however, that IMCO, in collaboration with other international organizations, 'should intensify its work' and that contracting states to the Public Law Convention should 'co-operate as appropriate in applying wholly or partially [its] provisions.'

The Public Law Convention authorizes states to take necessary measures on the high seas to prevent, mitigate, or eliminate 'grave and imminent danger to their coastline or related interests'[68] from oil pollution following a maritime casualty or acts related to such a casualty. It is doubtful whether this agreement adds anything to rights already possessed by states under the general international law doctrine of self-help,[69] but it is useful to have requirements for its lawful exercise reaffirmed and clarified in this particular context. The Private Law Convention, as its name indicates, establishes liability and provides for compensation (to a limit now around $17 million) to legal 'persons' (including states) who suffer damage from pollution resulting from the escape or discharge of oil from ships. This convention, however, applies exclusively to pollution damage 'on the territory including the territorial sea of a Contracting State.'[70] Finally, the 1971 International Convention on the Establishment of an International Fund for Compensation for Oil Pollution Damage, which grew out of the Brussels resolution on that subject, provides some supplementary compensation (up to somewhat more than double the earlier limit), but this is still limited by the 'territory including territorial sea'[71] (that is, narrowly defined exclusive interest) restriction. So none of these conventions is designed to afford protection to the high seas *per se* or to compensate for damage to inclusive interests in their protection and use.

Regionally, however, a bit more progress was made. The Agreement concerning Pollution of the North Sea by Oil[72] was signed at Bonn on 9 June 1969 by the governments of Belgium, Denmark, France, the Federal Republic of Germany, the Netherlands, Norway, Sweden, and the United Kingdom, and it entered into force on 9 August of that same year. As its name indicates, the agreement covers only pollution by oil, but otherwise the scope of its protection and the enforcement measures it describes represent an advance. The Bonn Agreement is to apply whenever the presence or prospective presence of oil within the North Sea area presents a grave and imminent danger 'to the coast or related interests of one *or more* Contracting Parties';[73] if read widely, this provision could be understood to encompass the inclusive interest of all these states in protection of the sea itself and its living resources against the effects of pollution. Furthermore, for the sole purpose of enforcing this agreement, the North Sea is divided into zones – not of sovereignty or jurisdiction – within which the contracting party or parties responsible are to make assessments of casualties and take certain ameliorative actions.[74] And shortly after the North Sea Agreement, Canada and Britain each passed legislation providing for similar anti-pollution regimes in zones off their coasts and outside their territorial waters.[75] In all these cases, active measures are envisioned for the prevention or abatement of pollution of large regions of the commons – such protection being somewhat auxiliary to protection of exclusive interests, but protection nevertheless.

The movement for international legal protection against marine pollution gained impetus in the context of the preparatory process for the 1972 Stockholm Conference and the conference itself. The pre-conference Intergovernmental Working Group on Marine Pollution, at its second session in Ottawa in 1971, considered the question of general principles of state responsibility to prevent and control activities which cause damage to the marine environment and adopted a Statement of Objectives and twenty-three General Principles for the Assessment and Control of Marine Pollution.[76] A major new feature of these formulations was the recognition that coastal nations 'have a particular interest in the management of coastal area resources' and that 'measures to prevent and control marine pollution must be regarded as an essential element in ... management of the oceans and their resources'[77] – thus opening the way for coastal states to assume a larger responsibility for protection against marine pollution. Three further principles submitted by the delegation of Canada on the rights and duties of coastal states in the exercise of such responsibility, however, although enjoying considerable support, failed to secure adoption in this forum.[78]

Meanwhile, more concrete success was achieved in the regional context. At the Conference on Marine Pollution held at Oslo in October 1971, the states bordering the North-East Atlantic adopted the Convention for the Prevention of Marine Pollution by Dumping from Ships and Aircraft.[79] The Oslo Anti-Dumping Convention established an absolute prohibition against the dumping of certain highly toxic substances and regulated the dumping of other toxic substances in the region of the North Sea and North Atlantic Ocean.[80] No convention is airtight in its drafting, and interpretative and political conflicts are bound to occur (for example, emergency disposal of nerve gas),[81] but these black and grey lists of prohibitions and restrictions were impressive in their comprehensiveness. The provisions for implementation by national permit systems[82] nevertheless left something to be desired. Each contracting party undertook to ensure compliance by ships and aircraft not entitled to sovereign immunity 'a/ registered in its territory; b/ loading in its territory the substances and materials which are to be dumped; c/ believed to be engaged in dumping within its territorial sea'[83] – that is, a flag and port state regime.

The Stockholm Conference itself endorsed the Ottawa Statement of Objectives and General Principles as guiding concepts for future scheduled conferences and took note of and referred to the Law of the Sea Conference the other three Canadian principles.[84] It also advocated rapid conclusion, 'preferably before the end of 1972,' of a general international ocean dumping convention.[85] Furthermore, Principle 7 of the Human Environment Declaration and eight detailed recommendations of the Stockholm Action Plan dealt particularly with the topic of marine pollution.[86] The Declaration and Action Plan subsequently received the endorsement of the UN General Assembly, in a resolution where the vote was 112 in favour to none against, with twelve abstentions (most of the abstaining votes being cast by the Soviet Union and Eastern European states that boycotted the Stockholm Conference).[87]

Since Stockholm there have also been some major developments in the legal order for controlling injurious use of the oceans. The general international Convention on the Prevention of Marine Pollution by Dumping of Wastes and Other Matter[88] was concluded in London and opened for signature before the end of 1972. The London Ocean Dumping Convention models its provisions on the Oslo regional precedent, prescribing similar black and grey lists for all marine waters other than internal waters of states.[89] It does, however, add a significant innovation as far as application is concerned. By contrast to the Oslo Convention, the London Convention is enforceable – not only against vessels and aircraft in the territory or territorial waters or flying the flag of a contracting state, but also against those that are 'under its jurisdiction believed to be engaged in dumping.'[90] It is thus the first broadly multilateral convention specifically to be enforceable by coastal states; but, so as not to prejudice the results of the Law of the Sea Conference, it also provided that the 'nature and extent of the right and the responsibility of a coastal state to apply the Convention in a zone adjacent to its coast' are to be resolved at a meeting of the contracting parties after UNCLOS III.[91]

A second important multinational convention concluded since Stockholm deals with discharges from ships. The International Convention for the Prevention of Pollution from Ships ('MARPOL' Convention),[92] done in London at a conference sponsored by IMCO in November 1973, extends the 1954 IMCO

Pollution of the Seas by Oil Convention discussed earlier with the objective of achieving 'the complete elimination of intentional pollution of the marine environment by oil and other harmful substances and the minimization of accidental discharge of such substances.'[93] It seeks to accomplish this by the proclamation of a comprehensive and detailed series of international regulations on the construction and equipment of ships and port facilities and the control of pollution caused by the carriage and discharge of oil, by other noxious liquid substances carried in bulk, by harmful substances carried by sea in package forms, or in freight containers, portable tanks, or road and rail tank wagons, by sewage from ships and by garbage from ships. Beyond the generally applicable norms, the convention expressly allows for more stringent pollution control standards in certain internationally designated 'special areas.'[94] The application regime, however, is much less clear. According to article 3, the Pollution from Ships Convention applies to 'a/ ships entitled to fly the flag of a Party to the Convention' and also to 'b/ ships not entitled to fly the flag of a Party but which operate under the authority of a Party.' The latter presumably includes the authority of port states in their internal and probably territorial waters, but is otherwise ambiguous; and the ambiguity is further enhanced by the reference in article 4(2) to proceedings for a violation 'within the jurisdiction of any Party to the Convention' to be taken in accordance with the law of that party or referred to the flag state. It is expected that clarification will be provided at the Law of the Sea Conference or elsewhere of what all this means in terms of the nature and extent of coastal and port state anti-pollution enforcement competence over adjacent and more distant waters.

Once again, as far as post-Stockholm collective accord is concerned, regional measures are far in advance of those requiring much more extensive collaboration. Most of the states which concluded the Oslo Anti-Dumping Convention,[95] joined by some of their land-locked neighbours, have reached agreement on certain forms of pollution of the sea emanating from the land. The Convention for the Prevention of Marine Pollution from Land-Based Sources[96] was finalized at a conference on that subject held in Paris in February 1974. The convention is actually rather narrow in its scope, defining 'pollution from land-based sources' only as that which arrives in the marine environment:

i / through watercourses,

ii / from the coast, including introduction through underwater or other pipelines,

iii / from man-made structures placed under the jurisdiction of a Contracting Party within the limits of the area to which the present Convention applies.[97]

It thus omits pollution which reaches the oceans by way of the atmosphere – that is, the more than three-quarters of all ocean pollution originating from land. Yet it is still a vast improvement over pre-existing law. As to enforcement, the contracting parties agree to ensure compliance with the terms of the convention through appropriate national legislative and administrative measures and to provide information thereon.[98] As the earlier Oslo Convention had done,[99] the Paris Convention establishes a regional commission to exercise overall supervision of implementation;[100] and a resolution at the Paris Conference further provided that the two commissions should hold common meetings with the same representatives and share a common secretariat.[101]

The activity goes on. Other conventions, for example the new 1974 International Convention on the Safety of Life at Sea (the SOLAS Convention),[102] have definite environmental implications as far as controlling injurious use of resources is concerned. Regionally, under the auspices of UNEP, the countries surrounding the Mediterranean have recently agreed on a general framework convention for the protection of that sea and two related protocols (one on pollution from ships and the other on combatting pollution by oil and other harmful substances).[103] And, of course, international law continues to be moulded by the ongoing practice and evolving expectations of states in their regular interactions with each other.

Finally, the Third UN Conference on the Law of the Sea is now in session, and its Third Committee has had under specific consideration the subject of preservation of the marine environment (the results of which deliberations were discussed above, 25–7). Whether or not these negotiations succeed in producing a new 'umbrella treaty' on the oceans, states will continue to have a common interest in ensuring that development of ocean resources is not accompanied by degradation of the marine environment, its life-supporting systems, and its living resources. It remains to be seen to what degree they will effectuate this interest by formal agreement or otherwise.

Atmosphere and space
Efforts at agreement for controlling the injurious use of inclusive resources have not been limited to the oceans. In the area of the atmosphere and space, interdependencies and therefore common interests are in many ways more universal and less immediately economic. These conditioning factors have certain contradictory implications as far as collective action is concerned. As a whole, however, the trends in the area of space law are similar to the expectations already generated in oceans law. Protection against pollution and determination of liability and compensation for that and other types of damage have been the primary subjects of consideration. Governments have had as their paramount concern the prevention of injury to their own territory and

people directly, and measures for preventing injury to the common resources themselves have been subordinated to this aim. By and large, as would naturally be expected, international law concerning use of the atmosphere and space is less developed than the law of the sea.

The UN Declaration of Legal Principles Governing the Activities of States in the Exploration and Use of Outer Space was the first international statement explicitly recognizing that '[s]tates bear international responsibility for national activities in outer space, whether carried on by governmental agencies or by non-governmental entities.'[104] But the declaration, which was adopted unanimously by the General Assembly on 13 December 1963, made no attempt to delineate the nature and scope of such state responsibility. That effort was subsequently undertaken in the negotiation of the Treaty on Principles Governing the Activities of States in the Exploration and Use of Outer Space, Including the Moon and Other Celestial Bodies, which was opened for signature on 21 January 1967 and entered into force on 10 October of the same year.[105] The Outer Space Treaty reiterates the international responsibility provision of the declaration,[106] and it further specifies that each state party launching or procuring the launching of an object into space and each state party from whose territory or facility an object is launched 'is internationally liable for damage to another State Party to the Treaty or to its natural or juridical persons by such object or its component parts on the Earth, in air space or in outer space.'[107] In regard to controlling particular injurious uses of the resources of space, the treaty prohibits the placing into orbit, installing on celestial bodies, or stationing in outer space in any other manner of any objects carrying 'nuclear weapons or any other kinds of weapons of mass destruction';[108] it also stipulates that states shall conduct studies and exploration of outer space and celestial bodies 'so as to avoid their harmful contamination and also adverse changes in the environment of the Earth resulting from the introduction of extraterrestrial matter.'[109] As far as enforcement is concerned, the contracting state on whose registry an object launched into space is carried is to retain jurisdiction and control over such object and any personnel thereof when in outer space or on a celestial body, and such objects are to be returned when found beyond the limits of the state of registry.[110]

As with uses of the oceans, the question of liability and compensation for damage caused by space objects has commanded considerable international attention. There have been both a convention and an important General Assembly resolution on this subject.[111] The Convention on International Liability for Damage Caused by Space Objects, which entered into force on 1 September 1972, sets up an interesting double standard of absolute liability for damage on this planet and a fault regime for extraterrestrial injury:

Article 2:
A launching State shall be absolutely liable to pay compensation for damage caused by its space object on the surface of the earth or to aircraft in flight.

Article 3:
In the event of damage being caused elsewhere than on the surface of the earth to a space object of one launching State or to persons or property on board such a space object by a space object of another launching State, the latter shall be liable only if the damage is due to its fault or the fault of persons for whom it is responsible.[112]

The common interests of all states in preventing and compensating for injury to the biosphere on which all human life depends are not directly served by these terms or by provisions made for their implementation. Claims are to be presented by the national state for its natural or juridical persons who suffer damage, or by another state in respect of damage sustained in its territory, or by a third state in respect of damage suffered by its permanent residents.[113] Compensation itself is to be fixed 'in accordance with international law and the principles of justice and equity.'[114]

Finally, while negotiated and viewed primarily as a disarmament measure, the 1963 Treaty Banning Nuclear Weapons Tests in the Atmosphere, in Outer Space and Under Water[115] also by its express terms represents an environmental protection agreement. The preamble to the treaty, in fact, states a double objective: '[s]eeking to achieve the discontinuance of all test explosions of nuclear weapons for all time ... and desiring to put an end to the contamination of man's environment by radioactive substances.' To these ends, the contracting states undertake to prohibit, to prevent and not to carry out any nuclear weapons test or any other nuclear explosion:

a / in the atmosphere; beyond its limits, including outer space; or underwater, including territorial waters or high seas; or
b / in any other environment if such explosion causes radioactive debris to be present outside the territorial limits of the State under whose jurisdiction or control such explosion is conducted.[116]

Significant additions to the area and terms of these multilateral prohibitions were later made by the Treaty on Non-Proliferation of Nuclear Weapons and the Treaty on the Prohibition of the Emplacement of Nuclear Weapons and Other Weapons of Mass Destruction on the Seabed and the Ocean Floor and in the Subsoil Thereof, and a regional agreement which should be mentioned in this regard is the Tlatelolco Treaty for the Prohibition of Nuclear Weapons in

Latin America.[117] Furthermore, on a related issue, the Convention on the Prohibition of the Development, Production and Stockpiling of Bacteriological (Biological) and Toxin Weapons and on Their Destruction,[118] which entered into force on 26 March 1975, is also obviously important from the point of view of creating a safe and healthy environment for present and future generations.

In general, the legal order for controlling injurious use of the atmosphere and other inclusive resources so as to avoid radioactive contamination of the environment is nevertheless far from complete. The Partial Test Ban Treaty, which entered into force on 10 October 1963, has yet to be signed by China and France, and those two states still insist on the right to carry out atmospheric nuclear tests. There is widespread international feeling that the atmospheric test ban has now been so generally accepted as to have become a norm of customary international law and therefore binding on all states, including non-signatories to the treaty. Some support for this contention is found in the Resolution on Nuclear Weapons Tests adopted at the UN Conference on the Human Environment; it considers the Partial Test Ban Treaty, the Seabed Denuclearization Treaty, and regional treaties such as the Tlatelolco Treaty and then resolves

a / To condemn nuclear weapons tests, especially those carried out in the atmosphere; b / To call upon those States intending to carry out nuclear weapons tests to abandon their plans to carry out such tests since they may lead to further contamination of the environment.[119]

But this formulation is at best hortatory, and the resolution was introduced hastily in the hope of using the influence and publicity of the Stockholm Conference to halt imminent French tests.

Australia and New Zealand tried to bring the environmentalist challenge to the legality of atmospheric nuclear testing before the International Court of Justice.[120] They both claimed that the carrying out of further atmospheric nuclear tests by France in the South Pacific Ocean would not be consistent with applicable rules of international law and would violate the rights of their citizens in common with the rest of the world to an environment free of radioactive fallout. In the course of the *Nuclear Tests* cases proceedings, the World Court issued Orders Concerning Interim Measures of Protection on 22 June 1973,[121] but France defied them by conducting a test over Mururoa just one month later and more tests a year later. The Court subsequently rendered judgment on 20 December 1974, holding that the dispute had disappeared and it was therefore not called upon to give a decision.[122] This was based upon the finding that the ultimate objective of both applicants was to obtain a termination of the atmospheric tests over the South Pacific and that, inasmuch as France had publicly announced the intention of ceasing the conduct of atmospheric tests upon completion of the 1974 series, the objective had in fact been obtained. Applications submitted by the government of Fiji to intervene in the two cases were therefore deemed to have lapsed and to require no further action.[123] France itself at all stages failed to appear before the tribunal, asserting its reservation of jurisdiction under Article 36 of the ICJ Statute. The vote was nine to six, with several separate opinions and dissents. A basic contention of the dissenters was that the objective of the applicants was too narrowly circumscribed by the Court, in that a declaratory judgment on the illegality of the tests was also sought and would have had value in and of itself.

Antarctica

Unlike the situation in the Arctic, Antarctic land masses and neighbouring waters were originally and have until now remained under inclusive competence and open for inclusive use on the basis of equality of opportunity. As early as 1959, largely at the instigation of the United States, those states maintaining substantial activities in the Antarctic held a conference to recognize and formalize their common interests. The resulting Antarctic Treaty,[124] to which originally twelve and now nineteen states are party, still has a few years to run. It does not constitute a renunciation of any previous rights or claims to territorial sovereignty in Antarctica, but it does provide that no acts or activities taking place while the treaty is in force shall constitute a basis for asserting, supporting, or denying any such claims.[125]

In any event, Antarctica is definitely an area of extra-special ecologic and particularly severe climatic conditions, whatever its political status. As regards controlling injurious use of the area, the most notable provision of the Antarctic Treaty is again the prohibition of radioactive pollution: 'Any nuclear explosions in Antarctica and the disposal there of radioactive waste shall be prohibited.'[126] And the ban in the Antarctic is more comprehensive than elsewhere: 'There shall be prohibited ... the testing of any type of weapons.'[127] The states party to the treaty also agreed to meet every year to consult together on matters of common interest, including the use of Antarctica for peaceful purposes only and the preservation and conservation of its living resources.[128]

Lately, however, Antarctica has increasingly become the focus of attention as a potential source of valuable resources. Primary attention is being paid to aquatic resources, especially krill, because of the vast quantities believed to exist and the supposed ease of their exploitation. There has also been greatly augmented interest in mineral resource activities. Consequently, the thirteen consultative parties under the Antarctic Treaty (the twelve original signatories – Argentina, Australia, Belgium, Chile, Denmark, France, Japan, New Zealand, Norway, the Union of Soviet Socialist Republics, the United Kingdom, and the United States – just joined by Poland) have been undertaking extensive efforts aimed at restructuring the existing regime. At their ninth regular meeting held in London from 19 September to 17 October 1977, they adopted Recommendation IX-2, which provides that a definitive regime for the conservation of Antarctic marine living resources should be concluded in 1978. In accordance with that recommendation, the 'Thirteen' held special meetings in London in November 1977 and in Canberra in February to March 1978, at which they produced a draft agreement. They scheduled another meeting for Buenos Aires in July 1978, for the purpose of completing revisions to their text. A fourth meeting, anticipated to be a diplomatic conference to conclude the agreement, is planned for Canberra around late 1978.[129]

International rivers

The problem of controlling injurious use of international rivers, other than boundary waters, is made somewhat more difficult by the privileged position of upstream states as far as pollution and other dangerous environmental consequences are concerned. On the other hand, due to proximity to ports and many other factors, lower states have tended to develop at a rate faster than their upstream neighbours, as well as being in a favourable position as regards access. Thus, the complementary facts that lower states need to ensure a supply of usable water from their upstream neighbours and upper states need to protect themselves in a variety of ways against monopolization of benefits by more advantaged users, as well as other political trade-offs, may produce favourable conditions for negotiation of controls of injurious uses of international rivers. Consequently, the trends in regard to co-operative agreement on uses of these inclusive resources do not differ radically from those of the oceans, space, and so forth from the environmental point of view. The standard of the 'community of interest of riparian states,' which was early developed to resolve situations where the inclusive use of international rivers conflicts with various exclusive activities, has been gradually broadened to include inclusive interests in pollution prevention and control.[130] Furthermore, the broader common interests of all states in oceans and seas have generated additional impetus for agreement on restricting pollution transmitted by rivers.

North American diplomats and scholars often proudly cite the Treaty Relating to Boundary Waters and Questions Arising along the Boundary between Canada and the United States[131] as one of the first examples – perhaps the first – of an internationally agreed duty not to pollute. There is a provision in the 1909 treaty that 'the waters herein defined as boundary waters and waters flowing across the boundary shall not be polluted on either side to the injury of health or property on the other.'[132] And as early as 1918, the International Joint Commission created by the Boundary Waters Treaty concluded that 'conditions exist which imperil the health and welfare of the citizens of both countries in contravention of the treaty.'[133] It is only very recently, however, that there have been active anti-pollution measures. In response to the warnings of a 1970 IJC report,[134] the governments of the United States and Canada managed to get together for co-operative action. The 1972 Agreement on Great Lakes Water Quality[135] sets forth specific water quality objectives and programs directed towards their achievement.

An interesting contrast to the Great Lakes–St Lawrence situation is the problem of the salinity of the lower Colorado River. The Colorado flows some 1300 miles from its headwaters in the State of Wyoming to its mouth in Baja California, Mexico; in its course it drains seven US states, is the border between the United States and Mexico for about twenty miles, and then drains the latter country for about one hundred miles before emptying into the Gulf of California.[136] A 1922 US interstate compact, in addition to regulating relations among the US riparians, recognized the possibility of a Mexican-American treaty apportioning the waters of the Colorado, and treaty negotiations did begin some time in the 1920s.[137] After a bitter and protracted struggle, the Waters Treaty of 1944[138] was finally concluded; it guaranteed Mexico a minimum apportionment of 1,500,000 acre-feet plus some surplus of the waters of the Colorado river,[139] but remained silent on the much-disputed issue of the quality of the waters. Late in 1961, the Mexican government sharply protested that the United States was delivering waters that were unusable (particularly because of their increased salinity), in violation of the rights of Mexico under the 1944 treaty and in derogation of US obligations under generally recognized principles of international law.[140] Much acrimonious debate followed, and it was not until 1973 – after over half a century of dispute and diplomatic negotiations – that the Agreement on the Permanent and Definitive Solution to the International Problem of the Salinity of the Colorado River[141] was reached. The new agreement not only stipulates a guaranteed annual volume, but also assures Mexico of a maximum average salinity of the waters crossing the border.[142]

In regard to other international rivers, the non-governmental International Law Association in 1966 adopted a model set of Rules on the Uses of Waters of an International Drainage Basin (the Helsinki Rules).[143] With respect to water pollution originating either within the territory of a state or outside its territory but caused by its conduct, article 10 establishes a strict regime of state responsibility:

Consistent with the principle of equitable utilization of the waters of an international drainage basin, a State:
a / must prevent any new form of water pollution or any increase in the degree of existing water pollution in an international drainage basin which would cause substantial injury in the territory of a co-basin State, and
b / should take all reasonable measures to abate existing water pollution in an international drainage basin to such an extent that no substantial damage is caused in the territory of a co-basin state.

This is backed up in article 11 by a stringent liability regime, providing that: in cases of violation of a/, 'the State responsible shall be required to cease the wrongful conduct and compensate the injured co-basin State for the injury that has been caused to it'; and in instances falling under b/, 'if a State fails to take reasonable measures, it shall be required promptly to enter into negotiations with the injured State with a view toward reaching a settlement equitable under the circumstances ...'

By comparison, under actual circumstances, governments represented in the Council of Europe have drawn up a draft European Convention for the Protection of International Watercourses against Pollution.[144] According to its article 3, states would similarly commit themselves to controlling pollution:

Each Contracting Party undertakes, with regard to international watercourses, to take:
a / all measures required to prevent new forms of water pollution or any increase in the degree of existing water pollution;
b / measures aiming at the gradual reduction of existing water pollution.

But on important related matters, article 21 declares only that the provisions of this convention 'shall not affect the rules applicable under general international law to any liability of states for damage caused by water pollution,' because after lengthy discussion 'it was finally agreed not to deal with the problem.'[145] The European draft convention (if it is ever finalized) will nevertheless be a step forward in so far as it seeks to protect these inclusive resources themselves in addition to the exclusive interests of particular states in the rivers, and it is consequently also more progressive from the point of view of controlling land-based sources of marine pollution. And, in general, international water quality law has been developing rapidly in recent years.[146]

Governments at the Stockholm Conference recognized the importance of rivers as sources of marine pollution as well as their intrinsic international importance. As a result, recommendation was made for a world registry of River Inputs into Ocean Systems (RIOS) and a World Registry of Clean Rivers.[147] No general international regime has, however, been agreed upon for controlling pollution from and of rivers. The regional Land-Based Sources Convention comes closest to serving as a precedent, since such sources are defined to include 'watercourses' and states pledge to implement the agreement 'jointly or individually as appropriate';[148] responsibility under these provisions would clearly encompass controlling emanations from international as well as national rivers. And at the Law of the Sea Conference, states have already managed to agree on a draft article which commits them, *inter alia*, to establishing national laws and regulations and also taking such other measures as may be necessary 'to prevent, reduce and control pollution of the marine environment from land-based sources including rivers, estuaries, pipelines and outfall structures, taking into account internationally agreed rules, standards and recommended practices and procedures'; the same draft article commits them, through competent international organizations or diplomatic conference, to endeavour to establish global and regional rules towards the same ends 'taking into account characteristic regional features, the economic capacity of developing States, and their need for economic development.'[149] The next step is specifying criteria and establishing the global and regional rules, standards, and recommended practices and procedures.

Customary international law rights of abatement
In addition to the above provisions concerning various inclusive resources, developed primarily through treaties and other international agreements (including by way of the UN 'legislative' process of resolutions, declarations, and so on), certain doctrines of customary international law are applicable for controlling the injurious use of inclusive resources. Besides their specific rights in contiguous zones or under statutory arrangements (for example, the IMCO Public Law Convention),[150] states have certain general rights of abatement beyond their territory or territorial waters. Under the varied doctrines of international self-help – including self-defence, self-preservation, and security – a state confronted with a major threat is permitted to exert the 'necessary and proportional' force to avert the danger or to abate its effects.[151] As well as allowing states to counteract or anticipate hostile operations coming directly from other states, these doctrines apply when the actual or potential danger comes from injurious use of inclusive resources.

The most famous example of the latter type of self-help in an environmental context is the behaviour of the United Kingdom in the face of the *Torrey Canyon* catastrophe.[152] The Royal Air Force bombed the stranded and damaged tanker in international waters in an attempt to halt further spillage of oil which was damaging the English coast. British plaintiffs were later, of course, still able to advance tort claims for damage which nevertheless resulted.[153]

The *Torrey Canyon*, the third largest tanker afloat at the time of the catastrophe, has been dwarfed by a new generation of supertankers. With the advent of deepwater ports, the possibilities of widespread pollution damage from accidents with these giants are enormous. Furthermore, the increase in number, size, and cargo capacity of bulk ore carriers and of nuclear-powered vessels also increases risks of calamity.[154] Some questions have been raised about the reach of customary international law doctrines of self-protection under these unprecedented circumstances, and this uncertainty has given rise to pressures for codifications such as the intervention provisions of the IMCO Public Law Convention.[155] There is, however, no question that states are not required to remain passive and await disaster at their shores, but may rather engage in some form of preventive or protective self-help.

Resources exclusively enjoyed
The regime for controlling the injurious use of exclusive resources is organized largely in terms of 'state responsibility'[156] for injury resulting from actions within national jurisdiction or control. To date this doctrine has been applied only narrowly for damage inflicted by one state or its nationals to the rights, property, or territory of another state, its nationals, or its activities. Principle 21 of the Stockholm Declaration, however, makes it clear not only that state responsibility extends to environmental responsibility, but also that it extends to inclusive resources – that is, 'damage to the environment of other states *or* of areas beyond the limits of national jurisdiction.'[157]

This consensus on the need to expand traditional doctrines of state responsibility to environmental damage including that beyond exclusive territorial limits was bolstered by General Assembly Resolution 2996, which declared that any resolutions adopted by the twenty-seventh General Assembly (and there were eleven on the human environment) could not affect Principles 21 and 22 of the Human Environment Declaration.[158] Finally, there is some support for the proposition that state responsibility reaches even further, beyond prevention of damage, to positive protection of the earth-space environment.[159] In any event, the law in this area is generally conceded to be in need of clarification and improvement. As companion to Principle 21 on the rights and responsibilities of states, Principle 22 of the Stockholm Declaration states the common conviction that

States shall co-operate to develop further the international law regarding liability and compensation for the victims of pollution and other environmental damage caused by activities within the jurisdiction of control of such States to areas beyond their jurisdiction.[160]

Such specific environmental law as now exists on controlling the injurious use of exclusive resources comes primarily from three decisions: the *Trail Smelter* arbitration,[161] the *Corfu Channel* case,[162] and the *Lac Lanoux* arbitration.[163] A fourth decision of direct interest, frequently overlooked by the publicists, is the *Gut Dam* arbitration.[164] Certain general principles of international law are, however, applicable by necessary implication in the light of the contemporary expectations of the world community, supplementary to this rather scanty caseload. In addition, some new conceptions of the rights and duties of states in this regard have begun to emerge.

Case law of state environmental responsibility
The international decisions can in general be said to indicate acceptance of a standard of strict liability for damage caused or deprivations resulting from manipulation of environmental variables.[165]

First, in the *Trail Smelter* arbitration,[166] Canada was held liable to the United States for damages for injury done in the State of Washington by fumes carried by the winds from a privately owned and operated smelter in British Columbia; Canada was also required to prevent such damage in the future. In essence, this case is an early reflection of the 'polluter pays' principle (which has since gained so much support), combined with injunctive relief.[167] At the time, the decision was something of an innovation at international law.

Pointing to the absence of international decisions dealing with air pollution, the tribunal observed that '[t]he nearest analogy is that of water pollution'; but again it found no international decisions directly on point.[168] It therefore felt the need to look to more general principles of international law and to particular state practice for guidance in resolution of the controversy. In a passage much quoted by environmentalists today, the tribunal reached its conclusion on the basis of a broad concept of state responsibility:

Under the principles of international law, as well as of the law of the United States, no State has the right to use or permit the use of its territory in such a manner as to cause injury by fumes in or to the territory of another or the properties or persons therein, when the case is of serious consequence and the injury is established by clear and convincing evidence ...

The tribunal holds that the Dominion of Canada is responsible in international law for the conduct of the Trail Smelter ... [I]t is the duty of the Government of the Dominion of Canada to see to it that this conduct should be in conformity with the obligation of the Dominion under international law as herein determined.

The Trail Smelter shall be required to refrain in the future from causing damage through fumes in the State of Washington. To avoid such damage the operations of the Smelter shall be subject to a regime or measure of control as provided in the present decision. Should such damage occur, indemnity to the United States shall be fixed in such manner as the governments acting under the Convention [the arbitration convention under which the tribunal was constituted] may agree upon.[169]

Secondly, in the *Corfu Channel* case,[170] the International Court of Justice had to decide the question of Albania's liability for failure to notify British ships about mines in the Albanian waters of the Corfu Channel. Having found such a duty to warn, the Court therefore held Albania responsible for the damage to the British warships when the mines exploded, and it further decided that Albanian sovereignty was not violated by the innocent passage of these vessels. The United Kingdom, however, was held to have no right to enter these Albanian waters in search of mine fragments, either as a means of preserving evidence regarding the earlier explosions or as a measure of self-help. In addition to basic tenets of sovereignty, the Court based its conclusion on 'certain general and well-recognized principles,' including 'every State's obligation not to allow knowingly its territory to be used for acts contrary to the rights of other States.'[171] It is this last observation which has been seized upon for its environmental law implications.

Thirdly, in *Lac Lanoux*,[172] the issue submitted to arbitration was whether a change proposed by France in its part of a river system would, if carried out without the prior agreement of Spain, constitute a violation of stated boundary treaties. The tribunal held for France, after finding that there would be full restitution of the diverted waters if the proposed hydroelectric project were carried out, and therefore that prior agreement was not required either under the treaties or under customary international law. *Lac Lanoux* has interesting procedural significance concerning the role of arbitration or adjudication in the development of international rivers and the settlement of attendant disputes.[173] From the substantive point of view, however, its chief contribution is the insistence by the tribunal that state responsibility on the basis of strict liability would have governed in the event of a contrary finding for Spain:

It would then have been argued that the works would bring about a definitive pollution of the waters of the Carol [the river through which the French Lake Lanoux empties into Spain] or that the returned water would have a chemical composition or a temperature or some other characteristics which could injure Spanish interests. Spain could then have claimed that her rights had been impaired.[174]

The three decisions above are those traditionally cited by the publicists as judicial precedents establishing state responsibility for environmental injury. The *Trail Smelter* arbitration is obviously a milestone in this development. While their relevance is apparent, the other two are somewhat far-fetched and reveal the lengths to which environmental lawyers have had to go in order to find sources in the field of international environmental law. The fourth, little-noted decision is perhaps of better than this marginal relevance.

Gut Dam[175] was another arbitration between Canada and the United States concerned with transboundary resource problems. The Canadian government built a dam between Adams Island in Canadian territory and Les Galops in the United States in 1903 and increased its height in 1904. In the 1965 arbitration, Canada was held liable to compensate for damages caused to US cottage owners through the consequent raising of Lake Ontario in 1951–2 – which obligation extended not only to the owners of Les Galops, 'but to any citizens of the United States.'[176] The government of Canada admitted its liability if the damage were attributable to the dam, and the tribunal found that it was so caused.

Some other traditional and emerging doctrines and practices
It has already been observed that in order to protect itself from damage a state may take measures to avert or abate imminent danger under traditional international law doctrines of self-help.[177] Lawful measures are logically more circumscribed when exercised against exclusive resources than when involving protective measures undertaken in inclusive resources, as might perhaps have been illustrated by the declaration of illegality of the British minesweeping operations in the *Corfu Channel* above.

Recently, international environmental law has begun developing beyond this time-honoured endorsement of limited avoidance measures and the *ex post facto* adjustment of damage claims to the creation of a positive duty to consult before undertaking activities which may damage the environment of other states. A formalized duty to provide proper environmental warning was proposed by the Working Group on the Declaration of the intergovernmental Preparatory Committee for the Stockholm Conference for inclusion in the Human Environment Declaration. The aborted 'draft Principle 20' would have read:

Relevant information must be supplied by States on activities or developments within their jurisdiction or under their control whenever they believe, or have reason to believe, that such information is needed to avoid the risk of significant adverse effects on the environment in areas beyond their national jurisdiction.[178]

Unfortunately, however, due directly to an ongoing environmental dispute between Brazil and Argentina over prior consultation responsibilities,[179] this principle failed to win acceptance at the conference. It was instead forwarded for consideration by the twenty-seventh UN General Assembly in the hope that a consensus would emerge by the next fall.[180] By that time, Brazil – the main opponent at Stockholm of draft Principle 20 – took the lead in presenting a modified resolution, which was co-sponsored by a large number of developing countries. It recognized state responsibility to provide 'technical data' with a view 'to avoiding significant harm that may occur in the human environment of the adjacent area'; but it further recognized that such data should not be misused 'to delay or impede the programmes and projects of exploration, exploitation and development of the natural resources of the States in whose territories such programmes and projects are carried out.'[181] A year later, the General Assembly passed another and stronger resolution on Co-operation in the Field of the Environment Concerning Natural Resources Shared by Two or More States, which stressed that such interstate co-operation 'must be developed on the basis of a system of information and prior consultation within the framework of the normal relations existing between them'; and it requested UNEP to report measures adopted in implementation of this and other provisions of the resolution.[182]

More recently, the Organization for Economic Co-operation and Development has adopted a recommendation on Principles Concerning Transfrontier Pollution[183] which includes a specific title on 'warning systems and incidents.' The basic duty to warn is stated as follows: 'Countries should promptly warn other potentially affected countries of any situation which may cause any sudden increase in the level of pollution in areas outside the country of origin of pollution, and take all appropriate steps to reduce the effects of any such sudden increase.'[184] This of course leaves out the question of gradual and cumulative effects which are nevertheless foreseeable, although they are partially taken care of by another title on 'principles of information and consultation' as to those works or undertakings 'which might create a significant risk of transfrontier pollution.'[185]

One area in which states have begun to show particular concern with warning and prior consultation commitments is that of weather and climate modification activities. As has already been reported, Recommendation 70 of the Stockholm Action Plan recommended that governments:

a / Carefully evaluate the likelihood and magnitude of climatic effects and disseminate their findings to the maximum extent feasible before embarking on such activities;
b / Consult fully other interested States when activities carrying a risk of such effects are being contemplated or implemented.[186]

To these ends, Canada and the United States have recently concluded an Agreement on the Exchange of Information on Weather Modification Activities. This treaty provides that information relating to weather modification activities of mutual interest acquired by a responsible agency of one party is to be transmitted to the responsible agency of the other,[187] and, more specifically that 'each Party agrees to notify and to fully inform the other concerning any weather modification activities of mutual interest conducted by it prior to the commencement of such activities ... as far in advance of such activities as may be possible.'[188] It further contains an explicit provision that '[t]he Parties agree to consult, at the request of either Party, regarding particular weather modification activities of mutual interest'; such consultations are to be initiated promptly and through telephonic or other rapid means of communication if necessary, and they are to be carried out in the light of the parties' laws, regulations, and administrative practices regarding weather modification.[189] Finally, there is a somewhat obtuse state responsibility article of qualification specifying that nothing in the new treaty 'relates to or shall be construed to affect the question of responsibility or liability for weather modification activities' or even 'to imply the existence of any generally applicable rule of international law.'[190]

Finally, states individually are beginning to accept responsibility for controlling injurious use of their resources through national laws and practices. These are not only self-protective, but also have international environmental ramifications. In the case of the United States, for example, section 102(C) of the National Environmental Policy Act (NEPA) of 1969 requires that a detailed environmental impact statement be included in every recommendation or report on proposals for legislation and other major federal actions 'significantly affecting the quality of the human environment'; in section 102(E) of the Act, Congress explicitly directs that federal agencies are to 'recognize the worldwide and long-term character of environmental problems.'[191] States collectively have also accepted an obligation modelled on this precedent in a draft article on environmental assessment negotiated at the current Law of the Sea Conference; it provides that states which 'have reasonable grounds for expecting that planned activities under their jurisdiction or control may cause substantial pollution of, or significant and harmful changes to, the marine environment' shall, as far as practicable, assess the potential environmental effects and communicate the results to UNEP or other international or regional organizations, which in turn should make them available to states.[192] Furthermore, aside from such assessment requirements, national substantive legislation itself can be drafted directly to take account of transnational or more broadly international environmental considerations. An illustration might be provided by the US Trans-Alaska Pipeline Act, which has provisions establishing strict liability for damages 'to any person or entity, public or private, *including* residents of Canada' that results from discharges of oil transported by vessels loading at terminal facilities of the pipeline.[193]

As far as the courts are concerned, in US legislation under the NEPA, a foreign citizen and foreign environmental organization were early permitted to intervene in order to ensure representation of their own interests.[194] And the Scandinavian countries have gone further and bound themselves by treaty reciprocally to recognize the standing of each other's citizens on the same basis as their own nationals to litigate environmental actions in their courts.[195] Also, there has been some movement for but not yet achievement of widely accepted uniform laws and reciprocal arrangements as to standing in national courts and administrative tribunals of domestic and foreign citizens claiming environmental injury or asserting environmental protection claims.[196]

FACILITATING CONSTRUCTIVE USE

Facilitating constructive use of resources is the positive analogue of controlling injurious use. As has already been observed, environmental goals include not only the minimization of injury but the realization of potential gains, although the two are in large measure indistinguishable.

Resources inclusively enjoyed

Mankind, according to the UN Declaration on the Human Environment, has the solemn responsibility 'to protect *and* improve the environment for present and future generations.'[197] In addition therefore to their common interest in controlling injurious use, nations have recognized their common interest in facilitating constructive or productive and harmonious use of environmental resources. But while sporadic or continuing crises have precipitated collective action towards the former goal of reducing negative externalities, common interests in the positive aspects of environmental management have achieved a lesser degree of realization. This suboptimalization of behaviour is more pronounced as regards inclusive resources than with exclusive resources (where there is obviously much more direct national self-interest in improving conditions).

Facilitating productive use of inclusive resources has primarily been a matter of conservation of renewable resources; under conditions of resource scarcities, questions of priority of users and apportionment of yields have had to be dealt with concomitantly. Increasingly, however, nations have also accorded consideration to organizing the productive and otherwise advantageous use of non-renewable inclusive resources. Facilitating harmonious use of inclusive resources in general has required clarification of principles of jurisdiction and limitations thereon and acceptance of international operational standards or 'rules of the road' by which to regulate and orchestrate behaviour.

Conservation of renewable resources
International conservation problems range along a continuum from those species which are in imminent danger of extinction to those which are suffering from milder and more reversible depletion or other unwise use. There are, of course, certain minimum reproductive levels beneath which renewable resources should not be permitted to drop. Beyond that, theoretical goals of conservation programs are frequently set forth in terms of 'maximum sustainable yield,' but that often represents an unrealistic or undersirable objective; instead, the preferable policy is likely to be some kind of 'optimum sustainable yield' which is determined by economic and social as well as biological criteria.[198] There has been much learned discourse about the meaning and relation of these two terms in the legal and political literature.[199] For present purposes, however, the central point is that when all the primary, secondary, and administrative costs of obtaining the maximum sustainable yield of any one species are taken into account, all users and/or others may be somewhat better off with acceptance of a somewhat lesser catch. What all this means, in essence, is that facilitating conservation – like its negative analogue of controlling pollution – is a threshhold concept and is affected by multiple ecological and political variables as regards definition of common interests as well as organization of collective action.

The most important international conservation concern has been for the living resources of the seas. A quarter of a century ago, states started concluding many regional and bilateral agreements relating to fisheries. These agreements seek to maintain the size of the stocks both by restricting individual activities of fishermen and by regulating fishing operations generally. The 1946 Convention for the Regulation of the Meshes of Fishing Nets and the Size Limits of Fish (which has since been amended several times),[200] as its name indicates, is an example of the former. The North-East Atlantic Fisheries Convention of 1959 aims at the latter type of regulation, setting up a regional fisheries commission mandated to propose such measures as establishing open and closed seasons, closing spawning areas to fishing, establishing size limits for any species, prescribing fishing gear and appliances the use of which is prohibited, and prescribing an overall catch limit for any species of fish.[201] And there are many similar conventions. At the beginning of the present round of UNCLOS negotiations, the compilers of a book of documents relevant to contemporary law of the sea negotiations listed no fewer than eighteen bilateral and plurilateral agreements (with multiple amendments) relating to Atlantic fisheries, sixteen such agreements relating to those of the Pacific, and fourteen other agreements relating to fishing and fisheries conservation.[202]

As far as broadly multilateral efforts are concerned, the legal landmark is the Convention on Fishing and Conservation of the Living Resources of the High Seas, which was one of the four agreements concluded at the 1958 Geneva Conference on the Law of the Sea.[203] The main substance of the Geneva Conservation Convention is contained in two of its initial articles:

Article 3:
A State whose nationals are engaged in fishing any stock or stocks of fish or other living marine resources in any areas of the high seas where the nationals of other States are not thus engaged shall adopt, for its own nationals, measures in that area when necessary for the purpose of the conservation of the living resources affected.

Article 4:
1 / If the nationals of two or more States are engaged in fishing the same stock or stocks of fish or other living marine resources in any area or areas of the high seas, these States shall, at the request of any of them, enter into negotiations with a view to prescribing by agreement for their nationals the necessary measures for the conservation of the living resources affected.
2 / If the States concerned do not reach agreement within twelve months, any of the parties may initiate the procedure contemplated by article 9 [dispute settlement by a special commission of five members].

This regime, however, must be interpreted in the light of the subsequent provision of the convention that '[a] coastal State has a special interest in the maintenance of the productivity of the living resources in any area of the high seas adjacent to its territorial sea,'[204] with the result that

any coastal State may, with a view to the maintenance of the productivity of the living resources of the sea, adopt unilateral measures of conservation apropriate to any stock of fish or other marine resources in any area of the high seas adjacent to its territorial sea, provided that negotiations to that effect with the other States concerned have not led to an agreement within six months.[205]

Thus, the special functional interests and rights which had long been claimed by coastal states in connection with certain aspects of inclusive resources were recognized by the international community rather early as regards conservation, although they have only very recently received any acknowledgement in relation to pollution prevention and control.

As to living resources in more imminent danger of extinction locally or totally, there are some long-standing international conservation treaties dealing with these species. In the case of marine mammals, the necessity for some limits on the kind and level of exploitation was early understood, and to a degree accepted, by those states whose nationals are the most substantial participants; and conservation efforts have been increasing as states have come to realize that the eleventh hour has arrived for whales, seals, and some other species. Fishery resources of certain common areas are also being seen to approach critical limits, and attempts are being made to do something about it. Implementation records of these conventions have been mixed.

An example of a regime which worked quite well for a long time is the 1911 Convention between the United States, Great Britain, Russia, and Japan for the Preservation and Protection of Fur Seals.[206] It prohibited pelagic sealing and sea ottering on the high seas and restricted such activity to areas within the control of the United States, Japan, and Russia. These countries were to share their catch from their own areas, and the United States (which got the great bulk of the seals under this arrangement) was also to make monetary payments.[207] The 1911 convention was successfully applied to fur seals until 1941, when Japan withdrew, and it was later replaced by the 1957 Interim Convention on Conservation of North Pacific Fur Seals[208] to which the United States, Canada, the Soviet Union, and Japan are the parties. The new convention is similar to the old but differs in two important respects: it adopts the goal of 'maximum sustainable productivity,'[209] which was not the stated purpose of its predecessor agreement; but it also qualifies this objective by taking account of the fact that seals may compete with other valuable resources for food and perhaps may use for their own food more valuable resources.[210]

Probably the most successful conservation measures are carried out by regional fisheries arrangements. An outstanding example of what was for about a quarter of a century a very workable and beneficial arrangement is the 1949 International Convention for the Northwest Atlantic Fisheries[211] and the International Commission for Northwest Atlantic Fisheries (ICNAF) set up thereunder. The convention itself provides for interstate panels for various subareas of its wide coverage, whose representatives are responsible for keeping under review the fisheries of their subarea and making recommendations for joint action by the contracting governments. Annually or semi-annually the ICNAF members, which originally numbered eleven and later expanded to sixteen states party,[212] get together for a massive bargaining session in which they apportion the allowable catch of each Northwest Atlantic species in three ways: according to total quota, by country, and by area. The Commission itself is supposed to perform functions of scientific investigation and statistical and other information gathering to assist in this regulatory effort.[213] Both the jurisdiction and procedures of ICNAF are now, of course, in the process of adjustment to take account of the advent of two-hundred-mile national fisheries zones in the area concerned.

By contrast, a well-known case where conservation measures have been blocked is that of whaling. Early agreements concluded by states in 1931 and 1937[214] were silent on the objectives of the regulation and placed no limits on the killing of whales, except to prohibit exploitation of certain species and of individuals below minimum sizes in exploitable species. The subsequent International Convention for the Regulation of Whaling,[215] signed in 1946, does include an explicit statement of conservation objectives and design some measures directed to these ends. The 1946 convention established an International Whaling Commission empowered to set up various regulations, including fixing the total maximum catch of whales by all parties together; the IWC is, however, specifically denied the authority to apportion this catch among states.[216] By the time of the Stockholm Conference, the international community had recognized the inadequacy of this regime and therefore recommended a ten-year moratorium on all commercial whaling, which recommendation has been repeated at subsequent sessions of the UNEP Governing Council; the moratorium was rapidly accepted by other whaling states but continued to be opposed by one hold-out, Japan (which alone accounted for more than 40 per cent of the whales officially reported caught), and controversies continue at annual meetings of the IWC and elsewhere.[217] As a result of such disagreement, not much progress has been made on this particular issue at UNCLOS III, which has only managed to agree on a draft article which currently reads as follows:

Nothing in the present Convention restricts the right of a coastal State or international organization, as appropriate, to prohibit, regulate and limit the exploitation of marine mammals. States shall co-operate either directly or through appropriate international organizations with a view to the protection and management of marine mammals.[218]

In the past few years there has been an upsurge of regional conservation conventions dealing with inclusive resources. Their diversity is illustrated by the following partial listing, all of which are treaties negotiated in the early 1970s. In regard to fisheries, seven states (including both East and West Germany) bordering on the Baltic concluded the Convention on Fishing and Conservation of the Living Resources in the Baltic Sea and the Belts and the Convention on the Protection of the Marine Environment of the Baltic Sea Area;[219] Iceland, Norway, and the USSR reached an Agreement on the Regulation of the Fishing of Atlanto-Scandian Herring;[220] Norway, the USSR, and the United Kingdom made a treaty limiting the total catch of North-East Arctic Cod;[221] and the United States and the USSR concluded an Agreement on Certain Fishery Problems on the High Seas in the Western Areas of the Middle Atlantic Ocean, which marks out fishery conservation zones on the high seas 'for the purpose of ensuring maximum sustainable yields and the maintenance of the said fisheries.'[222] As to other types of living resources, Brazil negotiated two separate agreements, one with the United States and the other with Trinidad and Tobago,[223] concerning the conservation of shrimp; Canada, Denmark, Norway, the United States, and the USSR, adopted an Agreement on the Conservation of Polar Bears;[224] and the signatories of the 1959 Antarctic Treaty added the Convention for the Conservation of Antarctic Seals.[225] From all of this diverse diplomatic activity, it is apparent that small groups of states that share particular immediate conservation interests are taking some collective measures in furtherance thereof in the absence of broad multilateral initiatives.

As far as broad-based international agreement is concerned, conservation of living resources is promoted by the 1973 Endangered Species Convention,[226] by which states have agreed to refrain from trade in specified species of flora and fauna in danger of extinction. Also, of course, conservation of marine living resources is a matter of great concern before the current law of the sea negotiations. To repeat an earlier observation, there has been a widely perceived need for new treaty provisions both in regard to international regulation of highly migratory species and to coastal state jurisdiction over fisheries in their exclusive zones.[227]

Meanwhile, some of these questions have been before the International Court of Justice in the *Fisheries Jurisdiction* cases.[228] The applicants, the

United Kingdom and the Federal Republic of Germany, made similar agreements with appellant Iceland in 1961 delimiting a twelve-mile zone for Iceland around the island.[229] In the beginning of 1972, however, the Icelandic Althing unanimously approved a Resolution on Fisheries Jurisdiction unilaterally extending its zone to fifty miles, claiming 'vital interests of the nation' and 'changed circumstances' and mentioning in particular the need for 'effective supervision of the fishstocks in the Iceland area';[230] Regulations Concerning the Fishery Limits off Iceland were subsequently promulgated, which prohibited all fishing activities by foreign vessels.[231] In consequence, Britain and Germany brought the matter to the World Court, asserting both violation of the prior agreements and that the action by Iceland was without foundation under international law. Iceland took no part in the proceedings. The Court issued and continued preliminary injunctions against Iceland and determined its jurisdiction to try both cases.[232] On the merits, by votes of ten to four, it held that the regulations were unopposable to the two specific countries protesting them, that Iceland was not entitled unilaterally to exclude the United Kingdom or the Federal Republic from areas between agreed fishery limits and the new fifty-mile limit or to impose restrictions on the activities of their vessels in that area, and that Iceland and the other two governments were under mutual obligations to negotiate in good faith for the equitable solution of their differences concerning their respective fishery rights.[233] The ICJ indicated its awareness of the movement for extension of fishery limits in the context of the Third UN Conference on the Law of the Sea, but it insisted that its judgment reflected the currently existing state of the law and that '[i]n the circumstances, the Court, as a court of law, cannot render judgment *sub specie legis ferendae*, or anticipate the law before the legislator has laid it down.'[234]

Advantageous use of non-renewable resources

International programs for securing the productive or otherwise advantageous use of non-renewable resources have primarily been concerned with developing a regime for the seabed beyond national jurisdiction. The world decision process has, however, also made some attempt to anticipate questions that will arise when it becomes technologically feasible to exploit non-renewable resources of space. And, as has already been observed, Antarctica is likely very soon to present such a challenge.

Plans and research and development activities have long been under way for mining the deep ocean floor, and quite recent developments suggest that certain such activities (in particular the mining of manganese nodules) may soon be economically viable as well as technologically feasible on a significant scale.[235] As a result, states have deemed it 'urgently necessary' to negotiate their differences and to create a regime governing resource exploration and exploitation beyond the seaward limits of coastal state jurisdiction over the continental shelf. Many delegations have submitted proposals in the course of the negotiations for the new Law of the Sea Treaty, but the differences in views have revolved around three critical issues: first, the nature of the re- source exploration and exploitation system – with alternatives of such activi- ties being conducted by states or their natural or juridical persons subject to regulation by an International Sea-Bed Authority, by the Authority itself either directly or through service contracts, or by some combination of the two; second, financial arrangements – the financial arrangements of the Authority itself and its operating arm, the Enterprise, and the financial terms of contracts for deep sea-bed exploitation; and third, the powers and functions of the Authority – voting arrangements, procedures for approval of regula- tions, dispute settlement procedures, and so on. By the seventh session of UNCLOS III in early 1978, these still remained as 'hard core' issues facing delegations, and Working Groups were set up to deal with each of them.[236] At long last, a great deal of progress on these issues appeared to be achieved. In the first area in particular, states had already agreed on the broad outlines of a dual 'parallel system' or 'banking system' under which the Enterprise and states and private companies will all exploit sea-bed resources; in Geneva in 1978, they managed to reach consensus or near consensus on several impor- tant subproblems: on production controls to protect land-based producers of nickel (based on a formula proposed jointly by the United States and Canada), on terms and conditions for transfer of technology to the Enterprise, and on several other issues. In the last two areas, there has been a very significant narrowing of the possibilities, and it is now beginning to seem more likely than at any prior stage in the negotiations that a new 'umbrella treaty' will actu- ally emerge from the negotiations.[237]

As regards facilitating productive and harmonious use of the resources of space and celestial bodies, the international community has not moved much beyond general provisions of use only for 'peaceful purposes.' There have been several UN General Assembly resolutions affirming this objective,[238] and the Declaration of Legal Principles and the Outer Space Treaty also recognize 'the common interest of all mankind in the progress of the exploration and use of outer space for peaceful purposes.'[239] Furthermore, just as there has been the UN Committee on the Peaceful Uses of the Sea-Bed and the Ocean Floor beyond the Limits of National Jurisdiction, so too there is the UN Committee on the Peaceful Uses of Outer Space. Since, however, states have agreed (at least for the present) that outer space, including the moon and other celestial bodies, is 'not subject to national appropriation ... by any ... means,'[240] they will undoubtedly in the future have to face the problem of deciding on a regime for the minerals and other inclusive resources of space – just as they are now doing for the sea-bed.

Regulation of transportation and communication uses
Besides facilitating constructive use of renewable and non-renewable or 'flow' and 'stock' resources, states have long agreed to measures for facilitating the harmonious enjoyment of transportation and communications uses – that is, of 'spacial extension' resources.[241] Towards this end, they have adopted certain principles of jurisdiction and 'rules of the road' for the oceans, international rivers, international airspace, and outer space.

The jurisdictional regime is basically the same for all of these media. All states are accorded equal opportunity to use the resources, and no state may assert jurisdiction over the vehicles, enterprises, and nationals of another state in these inclusive resources, except for certain functional purposes (as already discussed). Vessels,[242] aircraft,[243] and spacecraft[244] travelling beyond national boundaries are assimilated to their territory of registry, except where there is concurrent jurisdiction in foreign national waters or territory. Nation-states thus have the right to regulate the activities of their craft of registry and the responsibility to see that they follow international operational standards and other provisions of international law wherever they may journey.

As far as the 'rules of the road' themselves are concerned, the law of admiralty early began developing rules of navigation both by custom and by statute, and international navigation regulations covering the international carriage of goods and passengers are continuously being expanded and refined. For example, the duties and conduct of vessels are regulated by four sets of rules in the United States: the International Rules, the Great Lakes Rules, the Western River Rules, and the Inland Rules – the first three of which regulate international transportation.[245] In addition, there are all sorts of agreements on the safety of life at sea, salvage, bills of lading, friendship and navigation, and a multiplicity of ancillary issues.[246] Finally, as far as communications purposes of the oceans are concerned, there is not a wealth of highly specific treaty provisions, but the Geneva Convention on the High Seas makes the basic provision for 'freedom to lay submarine cables,' and there have been measures to protect this right.[247]

There have also been a host of international treaties and other agreements on international airspace and outer space. Besides the basic Warsaw Convention and the two Chicago Conventions on air services transit and civil aviation,[248] there have been several other international agreements regulating aviation, up to the most recent ones on hijacking and aerial sabotage.[249] As regards space travel, analogous common interests in harmonious use of spatial extension resources are advanced by the Treaty on Principles Governing the Activities of States in the Exploration and Use of Outer Space and the Agreement on the Rescue of Astronauts, the Return of Astronauts, and the Return of Objects Launched into Outer Space.[250] In addition, there is the basic International Telecommunications Convention[251] organizing the inclusive regime around which a host of international regulations have devolved dealing with radio, telegraph, telephone, television, and other subjects. And there have also been several notable agreements relating to various aspects of co-operation in satellite communications, especially the agreement relating to INTELSAT.[252]

For present purposes, these agreements need not be discussed at length or analysed individually. The crucial points are their very existence and recognition of the fact that they represent successful efforts by nation-states to promote their common interests by facilitating harmony in this type of inclusive resource use.

Resources exclusively enjoyed
With respect to their own exclusive resources, many states have gone much farther than existing international norms in instituting positive programs of public action. The US Clean Air Act, Air Quality Act, and Clean Air Amendments,[253] which together aspire to set up a comprehensive system of air quality control, afford an impressive example in just one area. Not only are national governments declaring their environmental aspirations, but some of them are creating the infrastructure essential to achieve these goals. In the United States, the Council on Environmental Quality (CEQ) has a significant advisory role, and the Environmental Protection Agency (EPA) undertakes major regulatory functions. World-wide, every major industrial power (as well as many other governments) has passed a considerable amount of environmental legislation and has an environmental Ministry, department, or agency of some sort.[254]

But we must concentrate here on examining what states collectively rather than individually have done and are doing to facilitate constructive use of their exclusive resources. The Convention on International Trade in Endangered Species of Wild Flora and Fauna[255] is an example of states collectively agreeing to prohibit certain uses of their own exclusive resources in the name of common conservation interests. Similar self-restrictions were imposed in two other conventions: the Convention for the Protection of the World Cultural and Natural Heritage[256] and the Convention on Wetlands of

International Importance Especially as Waterfowl Habitat.[257] By means of the former, each contracting state recognizes that 'the duty of ensuring the identification, protection, conservation, presentation and transmission to future generations of the cultural and natural heritage ... situated on its territory belongs primarily to that State' and binds itself 'to do all it can to this end.'[258] The World Heritage Convention also establishes an Intergovernmental Committee for the Protection of the Cultural and Natural Heritage of Outstanding Universal Value (summarily called the World Heritage Committee), creates a trust fund for the same purposes (the World Heritage Fund), and sets forth conditions and arrangements for international assistance for preservation of the heritage.[259] The contracting parties to the latter, *inter alia*, agree to designate suitable wetlands within their territory for inclusion in a List of Wetlands of International Importance and then to 'formulate and implement their planning so as to promote the conservation of the wetlands included in the list.'[260] As the necessity arises, they also agree to convene future conferences on the conservation of wetlands and waterfowl.[261]

In sum, most states individually are now realizing that facilitating constructive use of their own exclusive resources is a most important challenge to public policy and have initiated some active measures towards this end. Governments collectively are becoming aware that this is also a matter of common interest and have taken at least token and minimal collective measures towards this end.

PLANNING AND DEVELOPMENT OF NEW USES
The considerations involved in the planning and development of new uses are not fundamentally different from those motivating better regulation of traditional activities. The aim remains to take more full account of environmental costs and benefits – controlling injurious uses or negative externalities and facilitating beneficial effects or positive externalities. With new uses, however, there is often a wider range of choice, due to the absence of historical preferences and vested interests imposing practical and precedential limitations. It is therefore most desirable to attempt to plan the development of resources comprehensively from the beginning, rather than having to make piecemeal and often repeated *ex post facto* adjustments. A regulatory regime based initially on sound environmental principles is vastly preferable to a 'fire brigade' approach after grave problems begin to flare up.

Resources inclusively enjoyed
Urban planners have long advocated more effective planning of the physical environment and services of communities and the subsequent local and re-

gional development of resources in accordance with these plans. Many of these initially localized considerations themselves have international implications (for example, the amount of land devoted to food production, land-based activities which pollute the atmosphere and oceans, weather and climate modification efforts), and certainly the aggregate problem of planning and developing the resources of the shared earth-space environment is a matter of utmost international concern.

The importance of planning with regard to the use of inclusive resources has begun to be recognized by the international community. Principle 14 of the Stockholm Declaration, for example, flatly asserts: 'Rational planning constitutes an essential tool for reconciling any conflict between the needs of development and the need to protect and improve the environment.'[262] Until quite recently, nevertheless, very little attention has been given to comprehensive international environmental planning, and even less has been accorded to ensuring subsequent development in keeping with the priorities and objectives thus determined. Basically, it is through this aspect of the social process by which the international community organizes the use of resources that the diverse claims of unilateralism, bilateralism, regionalism, and multilateralism are reconciled or integrated and common interests are fulfilled in both the short and the long term.

To say that planning and development activities have been inadequate in both reach and detail is not to assert that nothing valuable has been accomplished. There have in the past been attempts, with varying degrees of success, at shared river basins development. A good case study is Canadian-US co-operation in respect of the St Lawrence and the Great Lakes. The 1909 Boundary Waters Treaty,[263] a major first step in the area of resource use planning, *inter alia* enumerated an order of precedence among the various uses of the waters in question and established an International Joint Commission endowed with jurisdiction over all controversies involving their use, obstruction, and diversion and other matters.[264] Much later, the 1972 Great Lakes Water Quality Agreement[265] represents some beginning of active development of the resources in accordance with jointly determined environmental objectives. Its preamble states the underlying conviction

... that the best means to achieve improved water quality in the Great Lakes System is through the adoption of common objectives, the development and implementation of cooperative programs and other measures, and the assignment of special responsibilities and functions to the International Joint Commission;

and the articles themselves set forth agreement on general and specific water quality objectives and on joint institutions as well as separate measures for their achievement.[266] Similar, although even less advanced, patterns of collective or co-operative planning and development of shared international rivers can be found in the cases of the Rhine and the Danube and the activities of their respective commissions,[267] and in other cases as well.

In addition to shared rivers, states have come to realize the need for inclusive planning and development of the resources of the sea-bed beyond the limits of national jurisdiction. As has already been reported, the Third UN Conference on the Law of the Sea expects to set up an International Sea-Bed Authority to deal with these matters, but the scope of its mandate and the extent to which it will be able to carry out or oversee development activities are still partially unresolved.[268] Finally, governments at the UN Conference on the Human Environment did agree on an Action Plan containing many recommendations for action at the international level,[269] and the UN Environment Programme and other agencies have accordingly initiated some measures, mostly in the area of environmental assessment.[270] But the enterprise is at present far from qualifying as comprehensive, ongoing, inclusive planning and management of inclusive environmental resources.

Resources exclusively enjoyed

With respect to resources exclusively enjoyed, an example of an international planning and development effort is the program of financial and other assistance to drought-stricken countries in the Sudano-Sahelian region Africa. In 1973, the UN General Assembly adopted a resolution urging all member states (developed ones in particular) to take all measures necessary to help these countries,[271] and most of the UN financial institutions, Specialized Agencies, and other organs, agencies, and programs have inaugurated intensified efforts to deal with this regional problem.[272] The six countries of the region have themselves established a Permanent Inter-State Drought Control Committee for the Sahel (CILSS) as part of their 'common will to take a stand against the calamity.'[273] In addition, the general subject of land, water, and desertification and the particular situation in the Sahel are accorded high priority in the activities of the UN Environment Programme, and a general United Nations Conference on Desertification was held in Nairobi from 29 August to 9 September 1977.[274]

With respect to national efforts, all nations plan – some with more formal term projections and explicitly stated goals than others. A hopeful collective initiative to promote the better planning and development of national or exclusive resources was 'Habitat,' the United Nations Conference on Human Settlements, which was held in Vancouver from 31 May to 11 June 1976. Habitat passed a 'Declaration of Principles' which recognized that 'the circumstances of life for vast numbers of people in human settlements are unacceptable, particularly in developing countries,' and that unless concrete ameliorative action is taken at the national and international levels these conditions are likely to be further aggravated as a result of: inequitable economic growth; social, economic, ecological, and environmental deterioration; world population growth; uncontrolled urbanization; rural backwardness; rural dispersion; and involuntary migration.[275] The conference made recommendations for national action concerning settlement policies and strategies; settlement planning; shelter, infrastructure and services; land; public participation; and institutions and management. It also adopted several resolutions containing recommendations for international co-operation, dealing with programs for such co-operation, financial implications, living conditions of the Palestinians in occupied territories, regional and subregional meetings, and post-Habitat use of audio-visual materials. In order to further this latter set of aims, it was also decided that there should be an intergovernmental body for human settlements and a small international secretariat; the human settlements secretariat, like UNEP, is in Nairobi.[276]

Access of people to resources

The issue of the access of people to resources has two aspects: the collective and the individual. In the former, it is primarily an issue of the relationship between developing and developed countries, between the 'have-nots' and the 'haves.' In the latter, the issue is that of the individual's access to the benefits both of the exclusive resources of his or her particular national community and of the inclusive resources of wider transnational communities. The fundamental goal in both cases is fairness or social justice of some sort, which – although largely indefinable by any empirical methods – is recognizable at least in its absence. Yet both cases also involve very objective goals of reducing the costs of resource use by the global community, since it is apparent that the present accelerated population growth and resource consumption, intolerably unbalanced economic and social conditions, and gross inefficiencies and diseconomies in the processes of international law and organization cannot continue indefinitely or even for very long.

The environmental problem of access to resources is exceedingly difficult, and measures to deal with either of its aspects are likely to be self-defeating and/or to counteract recommendations in regard to the other. One reason for this is that, as the Human Environment Declaration proclaimed, '[m]an is both creature and moulder of his environment'[277] and there is thus a juxtaposition of two perspectives – human resources and human dignity. Another reason is that patterns of international resource distribution are largely the result of historical developments or accidents rather than of ecological conditioning factors. Finally, the concept of fairness or justice is itself a mixed goal, and, although it has a central, universal core of meaning, both its moral content and its concrete ramifications vary somewhat among different societies and over time. The elusiveness of the goal, nevertheless, cannot justify ignoring it. The right of all people in present and future generations to 'freedom, equality and adequate conditions of life in an environment that permits a life of dignity and well-being'[278] is, after all, what our concern with the human environment is about.

GROUP CLAIMS TO ENVIRONMENTAL RESOURCES

The current concern with the human environment has arisen at a time when the energies and efforts of all or most countries of the world remain primarily devoted to the goal of economic growth. Since the rise of the nation-state, national governments have been seeking economic goals and using economic strategies as a major part of their international relations. It is only natural that recently emerged nations should attempt to do the same and to challenge historical inequalities. It is also unsurprising that land-locked, shelf-locked, and otherwise disadvantaged states should attempt to assert claims of traditional rights and equitable considerations in their particular struggle against the carving up of the oceans by their geographically more fortunate neighbours. Properly understood, economic and environmental objectives are often compatible or can be rendered so; nevertheless, a point in history has been reached when national and other communities must shape their actions throughout the world with a more prudent care for their environmental consequences.

Development and environment

The developing countries, which contain more than 70 per cent of the world population, account for only 30 per cent of its income, and the gap continues to widen. Having considered these facts, they have used their numerical muscle to pressure the UN General Assembly into adopting without a vote a 'Declaration on the Establishment of a New Economic Order,' which proclaimed that '[i]t has proved impossible to achieve an even and balanced development of the international community under the existing international economic order,' with a set of principles on which the new order should be founded, including '[t]he broadest co-operation of all States ... based on equity, whereby the prevailing disparities in the world may be banished and prosperity secured for all';[279] they have also voted through the General Assembly a new 'Charter of Economic Rights and Duties of States' in accordance with the declaration.[280]

Yet elsewhere, largely at the instigation of advanced industrial states, the international community as a whole has also proclaimed that '[t]he protection and improvement of the human environment is a major issue which affects the well-being of peoples and economic development throughout the world.'[281] The Stockholm Declaration also called upon all governments and peoples 'to exert common efforts for the preservation and improvement of the human environment.'[282] The question here is how these expressed goals of the world community interact and interrelate within the parameters of contemporary ecological interdependencies, resource scarcities, and other conditioning factors.

In many ways, the recent upsurge of environmental concern is in direct support of development efforts. As the Human Environment Declaration itself summarized:

In the developing countries most of the environmental problems are caused by underdevelopment. Millions continue to live far below the minimum levels required for a decent human existence, deprived of adequate food and clothing, shelter and education, health and sanitation. Therefore, the developing countries must direct their efforts to development, bearing in mind their priorities and the need to safeguard and improve the environment. For the same purpose, the industrialized countries should make efforts to reduce the gap between themselves and the developing countries. In industrialized countries, environmental problems are generally related to industrialization and technological development.[283]

Furthermore, the Stockholm Conference adopted eight recommendations on the subject of 'development and environment,' which advocate such measures as promoting regional co-operation, not invoking environmental concerns as a pretext for discriminatory trade policies, identifying threats to exports that arise from environmental concerns and developing common international environmental standards on products of significance in foreign trade, reviewing the practical implications of environmental concerns in relation to the distribution of future industrial capacity, distributing environmental technology to developing countries, and integrating environmental considerations into the review and appraisal of the International Development Strategy for the Second Development Decade in such a way that aid to developing countries is not hampered.[284] In keeping with these original objectives and in response to subsequent requests by the UN General Assembly and the UNEP Governing Council, the UNEP Secretariat prepared reports on the environmental impact resulting from the irrational and wasteful use of natural resources; this work was in turn considered by an Intergovernmental Expert Group on Environment and Development and on Environmental Impact Arising from Uses of

Natural Resources.[285] Then too, as far as concrete assistance is concerned, the UN Environment Fund supports many projects of greatest interest to developing countries, and the new UN Habitat and Human Settlements Foundation will be of primary benefit to them.[286]

In other ways, international environmental concerns may work to the indirect advantage of developing countries. Basically, the argument is that the very fact that developing countries have less industrial development and are less polluted may make it easier for them to meet common international quality standards internally. This perspective has a rather strong flavour of the allowance of 'tax havens' in domestic law, but – at least in the short term – there may be sufficient justification for certain kinds of 'environmental havens' or 'pollution havens' under international law for the purpose of promoting development. Misunderstanding of the phenomenon, however, can lead to environmentally unsound and illogical results.

At the Law of the Sea Conference, for example, there has been a good deal of discussion of the effects of differing levels of economic development on the duty to combat marine pollution. Some developing countries have claimed that their economic resources and stages of development should be taken into account in the setting of international standards and the allowance of national standards for pollution from vessels, from sea-bed exploration and exploitation and from land-based sources. In particular, with regard to vessel-source pollution, they have argued that coastal states should not be allowed to set vessel construction and design standards in such a way as to prejudice the development of their nascent maritime fleets and their use of second-generation ships. As concerns land-based sources of pollution and pollution from sea-bed activities, developing countries have also contended that their economic capacity does not allow them to take the same measures as industrialized states. Yet the counterargument has been made that developing countries are as susceptible as any others to harm to living resources and human health and amenities and other adverse effects of pollution. Lowering standards for their coastal waters, it would follow, would leave them worse off without commensurate gains. As a partial compromise, the pattern that is emerging from UNCLOS III is as follows: to forbid coastal states to set construction, design, manning and equipment standards; to allow them to set national laws and regulations and to endeavour to reach global and regional agreements regarding dumping and pollution from sea-bed activities; and expressly to provide for a double standard 'taking into account ... the economic capacity of developing States' only in regard to the establishment of global and regional rules, standards, and recommended practices and procedures for pollution from land-based sources.[287]

In addition, of course, developing countries may learn from the environmental mistakes of the advanced industrial countries. In part, this means avoiding – in so far as practible – the more obvious mistakes and distortions that have characterized the patterns of economic development of industrial societies. But beyond that, concern with the environment provides a new dimension to the development concept itself. A new emphasis is being placed on social and cultural goals and the 'quality of life' as part of the development process,[288] and perhaps in this dimension the developing countries will be able to provide the world with creative innovations.

Finally, however, absolute resource scarcities and the need to make resources available to protect and preserve the environment do mean that, on a global scale, environmental priorities and objectives will necessarily limit economic growth objectives narrowly defined. It is in this area of conflict that the most serious questions of equitable considerations arise. The costs involved are not only those of safeguarding the common resources, but also those which may emanate from the inclusion by developing and other countries of environmental protection measures into their development planning. In the Stockholm Declaration, it has been acknowledged by the international community, at least in principle, that advanced industrial states, which have caused so many of the adverse environmental consequences precipitating the present crisis, should bear the latter costs as well as the primary share of the former.[289] Yet between the proclamation and the assistance there often falls a shadow.

Other group claims
The primary and most pressing equitable group claims are those by the two-thirds of the people of the world who suffer the 'pollution of poverty.' But this does not or should not mean that no other such claims are deserving of accommodation. Another group of countries which have certain hopes or expectations, founded to a great extent on historical freedoms, are land-locked, shelf-locked, and otherwise geographically disadvantaged states. As support has been growing for exclusive economic zones, special fisheries regimes, and other arrangements to the benefit of coastal states, these geographically less fortunate states have been trying to raise equitable claims to preferential access to and allocation of fish, minerals, scientific research, and so forth. On the basis of long custom and tradition and under any of the varying conceptualizations of the 'common heritage of mankind,' these claims – especially those by developing land-locked states – may have some basis for recognition. So far at UNCLOS III, there has been some recognition of the right of land-locked states to participate in the exploitation of the surplus of the living resources of the exclusive economic zones of adjoining coastal states, as well as of certain

additional rights for geographically disadvantaged developing coastal states or those with certain 'special geographical characteristics'; other than providing that they shall be determined 'through bilateral, subregional or regional agreements,' however, there has not been much explicit arrangement for actualization of these rights.[290]

INDIVIDUAL CLAIMS TO RESOURCES
In the light of millennia of separate decision-making and of the recent tremendous upsurge of nationalism, it is hard to penetrate the acutely self-conscious separate sovereignty of 150 or so national governments with any perspective that is either supra- or transnational. The environmental perspective is both. It is concerned with humanity as a whole and with 'basic human rights – even the right to life itself';[291] this necessarily subsumes solicitude for the individual rights to human dignity and well-being of all the people that together make up humanity. It is also vitally concerned, however, with resource scarcities and absolute limitations on human consumption of material resources; these equations are affected not only by population growth rates and trends, but also by differential developments in regard to the consumption of energy and other resources. Thus, the complex issue of individual claims access to resources must be looked at both in terms of rights to participation in national communities and consequently in enjoyment of exclusive and inclusive resources and in terms of responsibilities for controlling numbers of people absolutely and in particular societies and areas.

In general, states have scarcely considered doctrines of human rights as they relate to environmental principles. Principle 1 of the Declaration on the Human Environment does, however, contain a summary statement:

Man has the fundamental right to freedom, equality and adequate conditions of life in an environment of a quality that permits a life of dignity and well-being, and he bears a solemn responsibility to protect and improve the environment for present and future generations. In this respect, policies promoting or perpetuating *apartheid*, racial segregation, discrimination, colonial and other forms of oppression and foreign domination stand condemned and must be eliminated.[292]

Nationality and movement of peoples
Nationality is the vehicle by which an individual can advance claims to participation both in the exclusive resources of his or her particular national community and in the inclusive resources of wider communities.[293] States have historically been permitted to prescribe and apply highly restrictive policies in the granting or denial of nationality. This regime has frequently resulted in

'stateless persons,' who are politically impotent both in the national territory in which they find themselves and in international areas. In recent years, however, there has been some attempt at ameliorating historic attitudes through the development of the contemporary human rights program. The Universal Declaration of Human Rights provides, at a minimum, that 'everyone has the right to a nationality' and that 'no one shall be arbitrarily denied the right to change his nationality.'[294]

Yet logically, if the right to a nationality is to be a meaningful vehicle of individual freedom and self-expression, it must be accompanied by certain concommitant international rights: the right to return, the right to leave, and the right to stay in a country. Some degree of freedom of movement between national communities is required for effectuation of these rights. Traditionally, however, states have imposed severe limitations on international freedom of migration in the form of immigration quotas, travel and visa restrictions, and so on on the one hand, and expatriation, deportation, and similar deprivations on the other. Rigid restrictions have been rejected in recent delineations of human rights, for example, the Universal Declaration of Human Rights,[295] the International Covenant on Civil and Political Rights,[296] the European Convention on Human Rights,[297] and others.[298] And the 1974 World Population Conference again repeated the general entreaty 'that Governments and international organizations generally facilitate voluntary international movement.'[299] This plea is, however, still being disregarded by many nations.

Population
The matter of numbers and concentrations of people – the 'population' question – is fundamental and critical to international environmental policy-making. Population problems have to be approached both from the point of view of people as resources (human resources) and from that of people as moulders of their own environment and destiny (human dignity). This raises urgent and sometimes contradictory policy considerations: on the one hand, people are an important base of power on the international scene, and individual governments are exceptionally jealous of their sovereign prerogative to determine national population policies; but, on the other, there remains the inescapable fact that no environmental measures in any dimension can ultimately be effective if the demographic explosion is not checked.

The basic numbers game of the population explosion is familiar, and the statistics and projections need not be repeated here.[300] But there are many other factors besides pure numbers that have to be taken into account. Population implosion effects, such as overconcentrations of people in cities and the deterioration of contemporary urban life, add other aspects. Differential rates of per capita consumption of food and energy resources are, of course, crucial considerations in view of present world scarcities. And additionally, many pressing social issues – such as the status of women, race relations, social welfare considerations, and so forth – must be taken into account.

The UN Conference on the Human Environment did not discuss population questions. One principle of the Stockholm Declaration does acknowledge their importance:

Demographic policies, which are without prejudice to basic human rights and which are deemed appropriate by Governments concerned, should be applied in those regions where the rate of population growth or excessive population concentrations are likely to have adverse effects on the environment or development, or where low population density may prevent improvement of the human environment and impede development.[301]

This general statement was, however, unaccompanied by any concrete policy recommendations. Nevertheless, this lack of attention reflected not a lack of concern, but rather the fact that the World Population Conference was already scheduled to be convened two years later.

The two-week UN World Population Conference met in Bucharest from 19 to 30 August 1974.[302] It ended by approving a Plan of Action containing many specific proposals and several more general resolutions and recommendations. It set no national or international population goals in terms of numbers of people, but it did suggest that with proper policies the world's present population growth rate of 2 per cent could be substantially reduced 1985.[303] As far as underlying principles and objectives are concerned, the starting point was taken that '[t]he formulation and implementation of population policies is the sovereign right of each nation';[304] but the plan did at least make reference to the essential element of international responsibility:

The effect of national action or inaction in the fields of population, may, in certain circumstances, extend beyond national boundaries; such international implications are particularly evident with regard to aspects of morbidity, population concentration and international migration, but may also apply to other aspects of population concern.[305]

In spite of the conference's somewhat weak treatment of international criteria in general and of international environmental concerns in particular, governments present did declare it 'a major contribution to the efforts of the international community to find appropriate solutions to the problems of population.'[306] Viewed against the background of past history, this may not be

an unfair assessment. Until very recently population issues have been regarded as being virtually exclusively within the decision-making competence
of individual sovereign states and not a fit matter for co-operation. It is only
within the last decade or so that transnational co-operative efforts to control
population growth and distribution have been inaugurated at all, and they
were at first narrowly limited to development assistance in the field of birth
control technology. This conference, by contrast, made recommendations on
such broad subjects as population and socio-economic policies, promotion of
the status of women, the family and development, rural conditions, urbanization, the improvement of health services, internal and international migration,
research, education and the interrelationships between population, development, resources, and environment.[307] The hope is now that a common 'Spirit of
Bucharest,' like the Spirit of Stockholm, will develop to become an integral
part of the overall process of world decision-making and be translated into
individual and collective action.

4

Processes of international
environmental organization

How is the world organized to manage resources so as to effectuate the policies and priorities defined by international environmental law? This cannot be answered adequately by means of a traditional 'functionalist' approach to international organization divided into specific substantive areas (for example, health, meteorology, labour, maritime affairs, and so on), but rather requires a more sophisticated functional, transfunctional, or programmatic analysis in terms of the processes characteristic of any public order system which makes collective choices and undertakes positive action. The contemporary global community, in other words, like its constituent national communities, maintains a comprehensive process of authoritative decision in which elements of both authority and effective control are combined. The overall process so maintained is made up of all those decisions and actions which characterize the different authoritative decision-makers, specify and clarify basic community policies, establish appropriate structures of authority, allocate bases of power for enforcement and sanctioning, and secure the continuous performance of all the different types of decision functions necessary to making and administering community policies.

The varying phases of this world decision process and the complex inter-relationships involved can be investigated with reference to certain basic problems:

1 / How is environmental information obtained, processed, and disseminated, and how can this process be improved?

2 / What are the means available for promoting environmentally sound thinking and practices? And how can these means be used more effectively to mobilize environmentalist demands?

3 / How are such demands crystallized and projected as legitimate international expectations or, in other words, as international law? In the light of the

present state system, what is the scope for improvement of the law-making function?

4 / What are the means and processes by which international machinery can be activated to initiate measures towards the ends determined by such community expectations? Where are the weaknesses, and what can be done about them?

5 / By what processes is there effective follow-through, so that choices and actions concerning the utilization of resources are characterized by environmentally sound policies? What measures can be taken to improve the record of implementation of policy choices in concrete circumstances?

6 / What is the regime for terminating old practices and policies and disposing of valid expectations created by them? Are there better or more equitable ways to ameliorate the costs of change to a more environmentally sound world public order?

7 / How is information obtained and evaluated concerning the overall environmental decision process itself? And in what ways can this systematic appraisal or 'watchdog' function be better performed?

Finally, although not a distinct phase of the world decision process, a few observations are necessary on a crucial international organizational problem endemic to each and all of the above considerations:

7 / What about funding? Where do the funds come from for environmental action on the international plane, and how can more be obtained?

**Gathering, processing, and
disseminating environmental information**

In order to have effective environmental management based on scientific principles, it is indispensable that there be an adequate intelligence input into the decision-making system. The gathering, processing, and dissemination of environmental information is one of the major concerns of the United Nations Environment Programme. The Action Plan adopted at Stockholm provided for a comprehensive global environmental assessment program, Earthwatch.[1] The activities this comprehends fall under four basic subheadings – evaluation, research, monitoring, and information exchange – and a great number of Recommendations of the Human Environment Conference have provisions pertinent to these functions.

The first Governing Council of UNEP decided that high priority should be given to the Earthwatch program, beginning with the establishment of a Global Environmental Monitoring System (GEMS).[2] Following that decision, an Intergovernmental Meeting on Monitoring, held in Nairobi in February 1974, outlined the objectives, principles, programs, goals and general guidelines of GEMS and listed environmental variables to be monitored; and in accordance with this blueprint, the Executive Director of UNEP hopes to have the system 'operational ... with results available, evaluated and published' by 1982.[3] GEMS is being constituted as a co-ordinated effort of nation-states, UN agencies, and UNEP to ensure that data on environmental variables (such as pollutant levels and the state of living resources) are collected in an orderly and adequate manner for the purpose of providing governments with a quantitative picture of the state of the environment and of the natural and man-made global and regional trends undergone by critical environmental variables. Participation in the system is voluntary, but, at least as regards oceanic data, delegates at the Law of the Sea Conference have agreed on a draft article which obligates states to 'observe, measure, evaluate and analyse, by recognized methods, the risks or effects of pollution of the marine environment' and to 'provide at appropriate intervals ... reports [of the results obtained] to the competent international or regional organizations, which should make them available to all States.'[4]

Besides the gathering, processing, and dissemination of information under GEMS, UNEP is carrying out or sponsoring many research and evaluation efforts in each of its six priority subject areas: human settlements, human health, habitat, and well-being; land, water, and desertification; trade, economics, technology, and transfer of technology; oceans; conservation of nature, wildlife, and genetic resources; and energy. To take the Regional Seas Programme as an outstanding example, under the Action Plan associated with the Barcelona Convention on the Mediterranean, more than eighty laboratories in fifteen coastal states around that body of water are co-operating in a joint program of baseline studies, information gathering, assessment, monitoring, data exchange and other efforts under the auspices of UNEP.[5] In addition, of course, virtually every one of the multitude of government agencies, international organizations, transnational associations, and private groups and individuals claiming environmental competence have study, research, and evaluation projects of some kind. And even organizations not traditionally cognizant of environmental implications do so occasionally. Sometimes this is required, as for US federal agencies under the 'impact statements' requirement of the National Environmental Policy Act of 1969,[6] and at other times it is just a wise step towards better overall planning and management.

In the vital area of information exchange, the UN Environment Programme hopes to become the international nucleus through the establishment of the International Referral System for Sources of Environmental Information (IRS). The system aims at comprehensiveness, covering 'not only quantitative data on the state of the environment (the analysed results of monitoring surveys, etc.) but also assessed documentary information about techniques for the monitoring, evaluation and control of environmental factors and bibliographical data.'[7] Its purpose is threefold: to act as a mechanism for the establishment of contacts between sources and users of environmental information, to provide a focus for an interconnected network of sources holding specialized information, and to collect and catalogue environmental information not held elsewhere in readily accessible form. The idea is to make use of a network of national, regional, and sectoral (or subject) focal points, and the facilities will eventually range from a carefully indexed but simple looseleaf service to a fully computerized system. Its sources have to be extremely varied, and co-operative arrangements were early made, for example, with the following: the Intergovernmental Oceanographic Commission (IOC) Task Team on Inter-disciplinary and Inter-Organizational Data on Information Management and Referral (in collaboration with UNESCO, FAO, WMO, IMCO, and IAEA) with overall responsibility for oceans; the Environmental Law Information System of the International Union for the Conservation of Nature (IUCN), working with the FAO legislative library, on legislation; the ECE Intergovernmental Centre for Documentation on Housing and the Environment; and the IUCN on wildlife and national parks. A number of governments have agreed to provide national focal points. It is envisaged that the users will eventually be equally wide-ranging and that use of the network will stimulate increased communication and efficient co-operation among all parties concerned.[8] As with GEMS, it is hoped that the IRS – along with the third major element of 'Earthwatch,' the International Register of Potentially Toxic Chemicals (IRPTC) – will be in full operation by 1982.[9]

Frequently, however, even where it has already been gathered and processed, access to information is much more than a question of referral to appropriate sources. There must be capability both to understand and to utilize it. Moreover, if information vital to global or individual national environmental management is only or most readily obtainable within a particular country or region, political or other barriers may interfere with its transfer. Along with their general spirit of *détente*, the United States and the USSR have recognized that the exchange of environmental information and experience is a function that can transcend political differences. Accordingly, in their bilateral Agreement on Cooperation in the Field of Environmental Protection, they have agreed to conduct activities by the following means:

exchange of scientists, experts and research scholars;
organization of bilateral conferences, symposia, and meetings of experts;

exchange of scientific and technical information and documentation, and the results of research on environment;

joint development and implementation of programs and projects in the field of basic and applied sciences;

other forms of cooperation which may be agreed upon in the course of implementation of this Agreement.[10]

This agreement was subsequently taken note of and reaffirmed, and the institutional arrangements were elaborated upon by the same two parties in the General Agreement on Contacts, Exchanges, and Cooperation in the Fields of Science, Technology, Education, and Culture.[11] The United States has also undertaken similar mutual commitments with the Federal Republic of Germany in their Agreement on Cooperation in Environmental Affairs,[12] and such co-operative arrangements are being made throughout the world by many different pairs and groups of states.

Lastly, in the general area of information exchange, considerable interstate controversy is still aroused on the subject of prior notification of proposed activities.[13] As has already been discussed, the duty to provide official and public knowledge of technical data relating to probable environmental consequences of proposed activities and the overall obligation of information and prior consultation have been accepted by the UN General Assembly (albeit in very qualified and oblique form), but the content and means of such notification have yet to be agreed upon and institutionalized.[14] In accordance with General Assembly Resolution 3129, UNEP consulted governments and made recommendations for a draft code of conduct setting forth general principles and guidelines for the conduct of states in the conservation and harmonious exploitation of shared natural resources, including provisions for notification, exchange of information, consultations, and a centralized information register.[15] A UNEP intergovernmental working group was then established, which at its fifth meeting over a period of two years managed to adopt a comprehensive set of Draft Principles of Conduct in the Field of the Environment for the Guidance of States in the Conservation and Harmonious Utilization of Natural Resources Shared by Two or More States.[16]

Promotion of environmentally sound thinking and practices

Knowledge for its own sake is not enough. There must be promotion, recommendation, or advocacy of policy in accordance with scientific principles. This involves the formulation and propagation of demands and the mobilization of support for environmentally sound thinking and practices.

In general, this function is conceded to be rather inadequately performed on the international or transnational level. The Human Environment Conference itself was in some measure successful in developing the missing ecological perspective among effective elites and the world community as a whole, but the new international awareness of common interests is still fragile and tentative at best. The 'energy crisis,' for example, provided disturbing indications of the perspectives of governments, multinational corporations, the press, and all sorts of groups and institutions, not to mention ordinary citizens. As one eminent analyst observed in the wake of Stockholm, 'there are signs of an increasingly widespread tendency to consider one year as the "year of the environment" and the next as the "year of the energy crisis." '[17]

Internationally, there has been some organizational effort directed towards achieving and sustaining awareness of and capacity to understand and deal with common environmental interests. Principle 19 of the Stockholm Declaration recognized at least part of the problem:

Education in environmental matters for the younger generation as well as adults, giving due consideration to the underprivileged, is essential in order to broaden the basis for an enlightened opinion and responsible conduct by individuals, enterprises and communities in protecting and improving the environment in its full human dimension. It is also essential that mass media of communications avoid contributing to the deterioration of the environment, but, on the contrary, disseminate information of an educational nature, on the need to protect and improve the environment in order to enable man to develop in every respect.[18]

The Action Plan adopted at the Conference highlighted the eclectic function of 'education, training and public information' as a crucial supporting measure to all environmental assessment and management activities.[19] Projects and plans developed in these areas are multiple and varied, ranging from special publications and audio-visual material to evaluation of the impact of mass communications on environmental awareness and action and to symbolic celebration of World Environment Day each 5 June (the date of the opening of the 1972 Stockholm Conference). To provide an effective base for carrying out some of these efforts, UNEP has created a Revolving Fund for information activities.[20] And as to future initiatives, in order to identify selective major environmental issues facing the world at the global and, to a certain extent, at the regional level, a State of the Environment Report is submitted each year to the Governing Council by the Executive Director.[21]

Yet international organizations are essentially the servants of the governments which compose their governing bodies, and they generally find it difficult to arouse public opinion and mobilize support for preferred policies. The human environment is nevertheless not without its particular champions among national governments. Sweden was the prime mover in calling for the convening of the UN Conference on the Human Environment.[22] Since the promulgation of the Arctic Waters Pollution Prevention Act and many subsequent initiatives,[23] the government of Canada is widely regarded as rampantly environmentalist in its foreign policy. The United States was instrumental in stimulating negotiation of the Oslo and London Dumping Conventions, was the prime mover of the environmental assessment draft article at UNCLOS III,[24] and has undertaken some recent environmental initiatives at UNCLOS under the leadership of its present head of delegation, Ambassador-at-Large Elliot L. Richardson. With UNEP headquartered in Nairobi, Kenya has also taken a lead in mobilizing support, particularly among developing countries.[25]

As far as the negative side of the question is concerned, of course, no government would be expected to come out overtly against the preservation of the human environment for the benefit of present and future generations. Even the refusal of the USSR and its Eastern European friends to attend the Stockholm Conference was allegedly based solely on political criteria rather than any lack of environmental consciousness.[26] Competing 'national security' and other special interests, both real and imagined, can nevertheless be destructive of common environmental interests and often divert attention and resources from environmental activism. And apathy itself is no insignificant barrier.

Transnational associations, too, have taken up the cause and become involved in environmentalist promotion activities. The numerous non-governmental organizations represented at the Stockholm Conference pledged in their NGO Declaration, *inter alia*, separately to 'mobilize support for the Stockholm decisions' and together to 'mobiliz[e] joint pressure for environmental change.'[27] The intergovernmental Governing Council of UNEP has adopted in its rules of procedure a provision giving observer status to and allowing limited participation in its sessions by non-governmental organizations 'having an interest in the field of the environment.'[28] But these activities have shown a tendency to ebb and flow in response to non-environmental factors. At UNCLOS III two years later, for example, only a very few of the Stockholm NGOs were represented, and there was noticeably less of an attempt at joint propagation of environmentalist demands on behalf of the world citizenry. Anti-environmental pressures, of course, were well represented. So, in regard to direct advocacy in the context of intergovernmental negotiations, the trends are mixed.

Large audiences are nevertheless being reached by some of these transnational associations by other means. The Club of Rome, through its now famous study *The Limits to Growth* and its subsequent *Mankind at the Turning Point*,[29] has had a significant impact in sounding a warning and arousing concern over ecological limits and conditioning factors. Its various challengers have also been influential in calling attention to the critical nature of environmental problems and the need to give more attention to them.[30] The International Council of Scientific Unions, to give another example, has had similar effects by means of its environmental subgroups with acronyms of SCEP (Study of Critical Environmental Problems), SCOPE (Scientific Committee on Problems of the Environment), and SMIC (Study of Man's Impact on Climate).[31] The International Institute for Environmental Affairs anticipated many critical issues in advance of their broad international emergence in its 1971 study *World Energy, the Environment and Political Action;*[32] and its reincarnation, the International Institute for Environment and Development, in its report by Robert Hallman entitled *Towards an Environmentally Sound Law of the Sea*,[33] made specific recommendations to delegates to the Law of the Sea Conference and advocated general policy to wider audiences as well. Then too, organizations like the Sierra Club, Friends of the Earth, and other international networks have national affiliates carrying on promotional activities in domestic arenas. They are joined in this, of course, by other interested national groups such as Resources for the Future, Natural Resources Defense Council, the Aspen Institute for Humanistic Studies, and the Center for Law and Social Policy, to give a few well-known US examples. The point is not the burgeoning numbers and the variety of these groups, but rather the facts that they do exist, that they are developing impressive scholarly and other potential, and that they may promote environmentally sound thinking and practices at many levels of community participation.

Individual voices are also not unheard. Maurice F. Strong, who was Secretary-General of the UN Conference on the Human Environment and later become Executive Director of the UN Environment Programme, spent a great deal of his time flying around the world trying to sustain interest and mobilize further support for particular environmental projects and general environmentalist concerns. Former Maltese Ambassador Arvid Pardo, who was once so instrumental in calling for a new review by the international community of the law of the sea, still tries to be a spokesman of some of these concerns. And social and natural scientists who write books translating scientific principles into their policy implications – such as Ward and Dubos, the Sprouts, Falk, and several others[34] – thereby make a significant contribution to the formulation and propagation of community demands. Finally, the environment is everybody's common interest. While boy or girl scouts collecting bottles for recycling do not have tremendous effects on world resource problems *per se*, the indirect effects of such activities in evolving ecological awareness and a sense of personal commitment are not to be discounted.

International environmental law-making

International environmental law-making or the prescribing function involves the formulation and projection of policy as authoritative community expectation. Historically, the making of transnational law has gone forward by way of articulated multilateral and bilateral agreement and unarticulated customary behaviour. Such law-making is accomplished by a whole range of means from diplomatic negotiations to UN and other parliamentary-diplomatic 'legislative' processes and to national legislative, executive, and judicial action. International environmental law is no exception, although there has been an unusually great amount of confusion and dispute over the validity of 'state practice' or 'unilateral action' in this field.

First of all, recent achievements in international environmental law-making by way of formal agreement among states are impressive. Within the area of marine pollution prevention and control alone, there have been many examples, as discussed in the preceding chapter, including: the International Convention for the Prevention of Pollution of the Seas by Oil (1954 with amendments up to 1971), the two 1969 IMCO conventions on oil pollution casualties and damage and its 1971 convention establishing an international oil pollution fund, the 1969 Bonn Agreement for Cooperation in Dealing with Pollution of the North Sea by Oil, the General Principles on marine pollution and the Statement of Objectives on the same subject (1971), the 1972 Oslo and London ocean dumping conventions, the International Convention for the Prevention of Pollution from Ships a year later, and the 1974 Paris Convention for the Prevention of Marine Pollution from Land-Based Sources.[35] Some of these agreements have not yet officially entered into force, but they are nevertheless highly significant in representing crystallizations of authoritative community expectations. Since they constitute the most prominent international code of behaviour, some states are already beginning to conform their activities to these accepted norms.

Diplomatic negotiations, of course, have been neither confined to the law of the sea nor restricted to matters of pollution and controlling other injurious use of resources. The Test Ban Treaty (1963), the Outer Space Treaty (1967), and the Convention on Liability for Damage Caused by Space Objects

(1971)[36] all embody prescriptions with relevance to the environmental protection of the atmosphere and biosphere. The World Heritage and Endangered Species Conventions (1973), the Gdansk Convention on Fishing and Conservation of the Living Resources in the Baltic Sea and the Belts (1973), the Agreement on Conservation of Polar Bears (1973),[37] and a variety of other multilateral and bilateral agreements have been addressed to the positive goal of the wise use and preservation of environmental resources for future generations. The central point is that all these articulations of conceptions of legal principles show both awareness of international environmental problems and a germ of willingness to do something about them.

Secondly, there has been an upswell of movement by the UN 'legislative' processes of resolutions, declarations, and conventions. The UN Declaration on the Human Environment and the Stockholm Action Plan have already been referred to; the Human Environment Conference also passed a resolution on institutional arrangements and several other resolutions.[38] The World Population Conference followed the same basic format for specification of community expectations.[39] At a more formalized level, UNCLOS III is working on its new 'umbrella treaty' to govern the future world order of the oceans. And a series of international conferences and meetings (on primary products, world population, food, water, desertification, human settlements, economic development, and so on) have to consider issues of existing or future international environmental law.

There are also United Nations resolutions. Right after Stockholm, the General Assembly passed no fewer than eleven resolutions having to do with the human environment, and it has kept passing more since.[40] Then too, the UNEP Governing Council passes a number of annual decisions or resolutions on the future development of UN activities.[41] Such formulations have the function of stamping authoritative community approval on international environmental policies and actions.

Thirdly, however, while there has always been some general scepticism about the value of international agreements of any kind and certain perennial challenges to the law-making competence of the UN General Assembly and other international organs have been made, there has been an especially pronounced nervousness about the role of national measures in the making of international environmental law.[42] It has been argued on the one hand that such unilateral action raises the potential problem of a 'patchwork quilt' of competing standards and criteria, and on the other that state practice continues to be a legitimate and essential means open to states for the progressive development of international law. Closer examination of the issues, however, reveals that there is no such stark dichotomy in fact and that it makes little sense to juxtapose 'unilateralism,' 'bilateralism,' 'regionalism,' and 'multilateralism' as competing alternative international approaches. International law-making proceeds simultaneously at all these levels, and the validity of particular outcomes must be judged not only by their compatibility with established norms and standards but also in the light of ecological conditions developing in advance of formalized legal processes.

When Canada passed its Arctic Waters Pollution Prevention Act and the accompanying Bill to Amend the Territorial Sea and Fishing Zones Act[43] in 1970, intense and heated debate by governments and among international lawyers ensued.[44] Given that many motivations and considerations other than environmental ones were involved, the United States in particular sharply criticized Canada for acting unilaterally. In what is probably the most blunt diplomatic note it has ever sent to its northern neighbour, the United States asserted:

International law provides no basis for these proposed unilateral extensions of jurisdictions on the high seas, and the United States can neither accept nor acquiesce in the assertion of such jurisdiction ...

We regret that the Canadian Government, while not excluding ... cooperative international approaches to our mutual problems involving the oceans, now proposes to take unilateral action to assert its own jurisdiction and establish its own rules pending the conclusion of international agreements satisfactory to it. For the reasons indicated earlier the United States can not accept these unilateral jurisdictional assertions and we have urged the Canadian Government to defer making them effective while cooperating in efforts promptly to reach internationally agreed solutions.[45]

Canada replied equally brusquely:

The Canadian Government is unable to accept the views of the United States Government concerning the Arctic Waters Pollution Prevention Bill and the amendments to the Territorial Sea and Fishing Zones Act, and regrets that the United States is not prepared to accept or acquiesce in them. The Canadian Government cannot accept in particular the view that international law provides no basis for the proposed measures ... Canada reserves to itself the same rights as the United States has asserted to determine for itself how best to protect its vital interests, including in particular its national security. It is the further view of the Canadian Government that a danger to the environment of a state constitutes a threat to its security. Thus the proposed Canadian Arctic Waters Pollution Prevention Legislation constitutes a lawful extension of a limited form of jurisdiction to meet particular dangers, and is of a different order from unilateral interferences with the freedom of the high seas such as, for example, the atomic tests carried out by the United States and other states which, however necessary they may be, have appropriated to their own use vast areas of the high seas and constituted grave perils to those who would wish to utilize such areas during the period of the test blast ...

It is well known that Canada takes second place to no nation in pressing for multilateral solutions to problems of international law, and that Canada has repeatedly and consistently shown its good faith by its continuous efforts to produce agreed rules of law. The Canadian Government is, however, determined to fulfil its fundamental responsibilities to the Canadian people and to the international community for the protection of Canada's offshore marine environment and its living resources, and the proposed legislation is directed to these ends.[46]

Some months later, in furtherance of this legislation, Canada created new fishing zones within 'fisheries closing lines' established across the entrances to bodies of water not enclosed within previous territorial sea baselines.[47] The United States once again protested vehemently:

The United States regards this unilateral act as totally without foundation in international law. The United States firmly opposes such unilateral extensions of jurisdiction and believes that outstanding issues concerning the oceans can only be resolved by effective international action;[48]

and this strong statement was reiterated and reaffirmed when the fishery closing lines became effective.[49] Canada nevertheless took a new initiative and passed amendments to the Canada Shipping Act extending its newly created fishing zones jurisdiction for the further purposes of preventing and controlling marine pollution.[50]

Yet, regardless of any such sharp exchanges, perceptions of environmental problems and positions of governments are not immutable. The United States itself soon adopted a piece of legislation seeking to protect its environment in respect of foreign vessels. Although the detailed Ports and Waterways Safety Act of 1972[51] does not reach geographically out into the high seas, it has as much potential effect over freedom of shipping and navigation and is as susceptible to charges, probably misplaced, of a 'patchwork quilt' effect. The Act inter alia allows the Secretary of the department in which the Coast Guard is operating to control vessel traffic by promulgating special standards for all vessels entering US ports, harbours, and other navigable waters. This includes the setting of standards with regard to matters of construction, design, equipment, and manning, as well as operation and movement of vessels.

Meanwhile the United States has an even more radical unilateral measure looming in the background of the Law of the Sea Conference. In 1971 the American mining lobby persuaded Senator Lee Metcalf of Montana to introduce a bill that would put the US government in the role of a nodule-mining agency in international waters, which bill has since been reintroduced and modified at subsequent sessions of Congress.[52] The United States would lease mining blocks to miners and even collect royalties on behalf of the international community. Furthermore, various versions of the bill also contain investment guarantees that if an uncongenial regime does take over in the international Area, the United States will be responsible for compensating its miners for their losses. While objectively such may not be the case, any number of Third World countries have at times flatly stated that passage of a US deep sea mining bill would be regarded as a decision to stake out a unilateral claim to mankind's common heritage and would wreck deliberations on all the widely ramified law of the sea issues under negotiation.[53] The controversy over sea-bed mining became particularly acute both within the United States and internationally in 1974 when Deepsea Ventures, a Delaware corporation, sent to the US Secretary of State a 'Notice of Discovery and Claim of Exclusive Mining Rights, and Request for Diplomatic Protection and Protection of Investments'; and the furore was not abated a few months later when the news broke that Howard Hughes' *Glomar Explorer* was mining for a sunken Soviet submarine instead of the manganese nodules it had allegedly been after.[54] Even now in late 1978, as UNCLOS III seems at long last to be approaching consensus on a regime for exploration and exploitation of the resources of the International Area, while the United States is still deliberating over national legislation, the controversy simmers on.[55]

In addition, the United States now has its own two-hundred-mile Fishery Conservation and Management Zone, whose promulgation has been closely related to the development of law at UNCLOS III. Senator Warren Magnuson of Washington and several co-sponsors in 1973 introduced a bill 'to extend on an interim basis the jurisdiction of the United States over certain ocean areas and fish in order to protect the domestic fishing industry, and for other purposes.'[56] After visiting the Caracas session of the Conference and becoming aware that it would certainly not produce a finished treaty, Alaska's Senator Ted Stevens, one of the co-sponsors of the legislation, renewed his advocacy of its rapid passage. He asserted the need for unilateral action until and unless multilateral solutions are finalized, and justified his contentions in terms of national self-interest in conservation:

I am more concerned now, than I was before going to Caracas, with the crisis threatening this Nation's fisheries, and am more convinced than ever that it is crucial to the overall best interests of the United States that Congress take immediate action to extend our fisheries management zone to 200 miles [from 12 miles] ...

I continue to believe that international law governing the oceans is attainable and eminently worthwhile, particularly law preserving world fisheries. But if the present lack of concern for basic conservation principles on the seas persists among some foreign fish operators for the indefinite amount of time necessary to conclude the law of the sea negotiations then we may be left very little of value to preserve. Enactment of S. 1988 [the fish bill] by the United States will not be prejudicial to the successful formulation of a law of the sea treaty. On the contrary, S. 1988 conforms to world trends.[57]

Subsequently, while UNCLOS III developed the concept of the exclusive economic zone, the United States Congress enacted the Fishery Conservation and Management Act of 1976, which prohibited foreign fishing without permits in a two-hundred mile zone off US coasts.[58]

Finally, to complement this extension of fisheries jurisdiction, Senator Edmund Muskie of Maine introduced a bill to establish a two-hundred-mile Marine Pollution Control Zone by amending the Federal Water Pollution Control Act and the Ports and Waterways Safety Act to extend US vessel pollution control this distance from US coasts.[59] It would thus be even more extensive in its offshore reach than the Canadian Arctic Waters Pollution Prevention Act of 1970 or even the 1971 Canada Shipping Act Amendments.[60] In advancing his proposals, Senator Muskie explained that, while fully supporting multilateral efforts, concerned governments like that of the United States must undertake environmental responsibilities for the protection of their offshore marine environment and living resources:

I had the privilege of attending the conference at Caracas as a member of the US delegation. I was disappointed to find that pollution control was and continues to be relegated to a low priority, and many countries, perhaps a majority, have made it clear that they do not desire strict standards for pollution control ...

Furthermore, even if a Law of the Sea Treaty were to establish an international regime, the United States ... would still want to have the authority and the jurisdiction to adopt stricter standards and additional measures for the ocean area surrounding its coastline.[61]

In somewhat modified form, the Muskie bill was recently enacted into law by the Clean Water Act of 1977; section 58 of that act extends US jurisdiction over vessels to prohibit discharges not only in the territorial sea and contiguous zone, but also

in connection with activities under the Outer Continental Shelf Lands Act or the Deepwater Port Act of 1974, or which may affect natural resources belonging to, appertaining to, or under the exclusive management authority of the United States (including resources under the Fishery Conservation and Management Act of 1976).[62]

The polemic of why and when to take national actions in the absence or in advance of multilateral agreement is, of course, not unique to the US-Canadian dialogue. Several other countries have legislated unilaterally concerning the environment of areas outside their national jurisdiction. The British Oil in Navigable Waters Act of 1971 is a notable case.[63] Other examples that have been cited for this same quality of extraterritorial reach include: the Australian Navigation Act of 1912–70; the Netherlands Pollution of Surface Water Act of 1969, together with its Implementation Decree of 5 November 1970; and the New Zealand Oil in Navigable Waters Act of 1965, together with its Regulations of 3 May 1971.[64] Even more comprehensively than these countries or the United States, Canada seems to have extended the reach of its offshore antipollution jurisdiction through its own now-two-hundred-mile fishing zones area;[65] while this jurisdiction appears to have been extended by operation of law, it has not yet been fully exercised in practice.[66]

The intention here is not to assess the international legality or utility of any of these particular pieces of legislation or of the various national executive and judicial actions towards similar ends. The point to note is, rather, that all of these measures were developed concurrently with preparations for and deliberations of the world-wide UN Law of the Sea Conference, which is undertaking consideration of much of the same subject-matter.[67] The making of international law, to repeat, goes forward concurrently and often with an impressive degree of complementarity of purpose and effect at multiple levels of decision-making.

Invocation of international machinery for environmental causes

With international law, as in any legal system, it is not always enough to have the letter of the law on one's side. There must often be some means of setting in motion machinery to interpret concrete circumstances in reference to the legal prescriptions. This role is usually performed by courts or arbitral tribunals, although various international and national administrative bodies may also be called upon to issue authoritative interpretations. Thus, in order to stimulate the application of the law or challenge its misapplication, advocates must have access to these arenas or be able to find a surrogate or champion with the appropriate standing. This general problem is especially acute with international environmental law, as the important question 'Who shall speak for the commons?' or even, more conservatively 'Who can speak for the commons?' remains largely unresolved.

Only states can bring cases before the International Court of Justice or most international arbitral tribunals, but they often do so in a representative capacity.[68] The overwhelming trend of international decision with regard to standing has been to permit the state of nationality – and only the state of nationality – to assert the public law claims of individuals and corporate entities. Furthermore, states are still regarded as having an option as to whether or not they will protect their nationals, since international law imposes no duty on a state to do so. This law has been developed in such famous decisions as the *Nottebohm* case (narrowly restricting the instances in which a state is entitled to exercise its protection of a naturalized person by means of international judicial proceedings),[69] the *Flegenheimer* arbitration (again limiting state discretion as to who is a national for purposes of invoking international machinery),[70] and the recent *Barcelona Traction* case (holding that the protection of a company and hence of its shareholders is entrusted to the state of incorporation and only to that state).[71] The last-mentioned decision is particularly significant because of its holding that 'the protection vested in the national State ... cannot be regarded as extinguished because it is not exercised,' and therefore that in instances of non-exercise of diplomatic protection there is no right or option of protection secondary to that of the national state.[72] In other words, a state is entitled to protect its nationals and their exclusive interests; but if the government in question does not choose to do so, there is clearly little redress left under general international law.

There has been some attempt to modify this rigid regime by treaty. The Convention on International Liability for Damage Caused by Space Objects, for example, contains a specific provision for secondary and tertiary protection:

1 / A State which suffers damage, or whose natural or juridical persons suffer damage, may present to a launching State a claim for compensation for such damage.

2 / If the State of nationality has not presented a claim, another State may, in respect of damage sustained in its territory by any natural or juridical person, present a claim to a launching State.

3 / If neither the State of nationality nor the State in whose territory the damage was sustained has presented a claim or notified its intention of presenting a claim, another State may, in respect of damage sustained by its permanent residents, present a claim to a launching State.[73]

And if the claim is not satisfied through diplomatic negotiations within one year, the Convention further specifies that the claimant state (in whatever representative capacity) and the launching state shall establish a claims commission at the request of either party.[74] Yet this is an exception to the prevailing rule and a rather highly qualified one at that; it has not yet been tested in an actual controversy.

As far as national tribunals are concerned, however, there is some indication of an emerging trend towards a widening of their portals in transnational environmental matters. When an incident has caused oil pollution damage in the territory (including territorial sea) of a particular state or states (wherever the escape or discharge responsible may have occurred), the International Convention on Civil Liability for Oil Pollution Damage stipulates that actions for compensation may only be brought in the courts of the damaged state(s); it does additionally, however, commit each contracting state to 'ensure that its Courts possess the necessary jurisdiction to entertain such actions for compensation.'[75] With regard to any kind of transnational environmental nuisance or damage, Denmark, Finland, Norway, and Sweden have provided for access of each other's nationals to their courts. More specifically, they have agreed that any person who is or may be affected by a nuisance caused by environmentally harmful activities or is seeking compensation for damage caused by such activities in another contracting state

shall have the right to bring before the appropriate Court or Administrative Authority of that State the question of the permissibility of such activities, including the question of measures to prevent damage, and to appeal against the decision of the Court or the Administrative Authority to the same extent and on the same terms as a legal entity of the State in which the activities are being carried out.[76]

Without an explicit international agreement, a Canadian citizen and a Canadian environmental organization were early permitted to intervene in litigation in US courts under the National Environmental Policy Act of 1969 to ensure representation of their own separate interests.[77]

But all of the above developments pertain to what are primarily exclusive interests. Who has standing to speak for inclusive common interests in the protection and preservation of the human environment has not yet been clearly established. In this connection it is significant that both appellants in the *Nuclear Tests* cases before the International Court of Justice claimed to be representing not only their own exclusive interests but also the compatible inclusive common interests of the entire international community. Australia asserted not only that the deposit of radioactive fall-out on its territory is in violation of Australia's own sovereignty and territorial integrity but also that

[t]he right of Australia and its people, in common with other States and their peoples, to be free from atmospheric nuclear weapons tests by any country is and will be violated ... [and that] the interference with ships and aircraft on the high seas and in the superjacent airspace, and the pollution of the high seas by radioactive fall-out, constitute infringements of the freedom of the high seas.[78]

New Zealand made analogous assertions, including that continuation of the testing 'violates the rights of all members of the international community, including New Zealand.'[79] The Court neither passed judgment nor commented on this point. It has, of course, been held in the *Reparations* case that international organizations may bring claims against governments on behalf of their collective membership (or of individual victims),[80] but the problem remains of under what conditions one state may appoint itself champion for the inclusive interests of the world community on a particular issue.

Application of policies in concrete circumstances

Even where there has been a policy decision and the law clearly applies or has been adjudged to apply, it does not automatically follow that the law will be implemented in practice. All governments and other actors have a common interest in the application or enforcement of international law in general and of international environmental law in particular, but they have not proceeded to develop a multilateral enforcement capability (except sometimes for certain highly circumscribed security situations). Given the difficulty of obtaining such concerted action, it has been found necessary to look elsewhere than at the central collective institutions for the major enforcement effort. Consequently, while there has been an impressive upsurge of multilateral initiatives in the prescription of international environmental law, the trend is clearly towards increased reliance on different individual states for the application of collectively derived prescriptions or, in their absence, of reasonable unilateral standards and criteria.[81] The main problem is to determine which states should be charged with the increased rights and duties of implementation. The issues involved have been most prominently manifested and debated in the controversy in the law of the sea negotiations over flag, port, and coastal state responsibility for the enforcement of marine environmental protection measures.

While the list of environmental agreements or those with implications for the protection of the marine environment is growing rapidly, little progress has been made in ensuring their implementation. The main defect of the treaties is that almost all of them rely only upon flag states for their application and make no alternative provisions for the enforcement of the standards established. The 1954 International Convention for the Prevention of Pollution of the Sea by Oil and its amendments up to 1971,[82] for example, regulate discharges on the high seas and prohibit such discharges within certain distances from land, but it has been pointed out that:

The conventions' effectiveness [that of 1954 and its amendments] was limited, however, since their enforcement lay [exclusively] within the jurisdiction of the states of registry. They contained no recognition of a coastal state's right of abatement, even in the defined 'prohibited zones.' Nor did they deal with the vexed issues of liability for harm.[83]

As a result, one authority reported, after the treaty had been in force for fifteen years, that only two prosecutions for a convention offence outside a state's territorial sea had ever been recorded.[84] Furthermore, the United States Coast Guard records show that of seven cases of discharges in violation of the convention which the United States referred to flag states during the period of 1969 to 1972, 'modest penalties were assessed in only 2 cases.'[85] Almost all subsequent international agreements – up to and including the 1969 and 1971 IMCO oil pollution conventions,[86] suffer from the same basic defect of sole reliance on flag state enforcement.

The duty of flag states to live up to their international obligations and enforce these agreements has naturally been and continues to be much reiterated. The General Principles for the Assessment and Control of Marine Pollution, for example, expressly provide:

All States should ensure that vessels under their registration comply with internationally agreed rules and standards relating to ship design and construction, operating procedures and other relevant factors. States should co-operate in the development of such rules, standards and procedures, in the appropriate international bodies.[87]

But the international phenomenon of 'flags of convenience,' where there may be only shadowy authority and control by the state of registration over its ships,[88] undermines the effectiveness of any legal regime critically dependent upon this obligation. And, in any event, most flag states have only a general interest in common with the entire international community in the implementation of marine environmental protection measures as regards their ships; they lack an independent incentive to undertake effective efforts towards this end. Consequently, the international order is beginning to move away from this admittedly inefficacious, unidimensional approach and towards reliance on the complementary roles of port and coastal states in the preservation of the marine environment.[89]

Port state jurisdiction for the prevention of vessel-source pollution is seen by some as a viable and effective supplement to flag state enforcement. This approach would entitle states to enforce the provisions of international legal obligations against foreign vessels found in their ports, perhaps irrespective of the area in which a violation occurred. As has already been described with reference to the prescription of law, a form of port state jurisdiction has already been incorporated in existing national legislation – namely, the US Ports and Waterways Safety Act.[90] The enforcement provisions of this act give the Secretary of the department in which the Coast Guard is operating broad investigatory powers in regard to any incident, accident, or act which involves damage to vessels, bridges, or other structures on or in the navigable waters of the United States or which 'affects or may affect' the safety or environmental quality of US ports, harbours, or navigable waters.[91] As part of this general authority to investigate, the Secretary may issue subpoenas for witnesses, documents, and other evidence and may request the Attorney General to invoke the aid of US district courts to compel compliance.[92] He is also empowered to assess and collect civil penalties up to $10,000 and criminal penalties of not less than $500 or more than $50,000 for inadvertent or wilful violations.[93] In the multilateral context, in addition to acceptance of the general concept of universal port state enforcement at UNCLOS III, the new Convention concerning Minimum Standards in Merchant Ships drafted under the auspices of the International Labour Organization is to be enforceable by a state 'in whose port a ship calls in the normal course of its business or for operational reasons.'[94]

Others see coastal state enforcement jurisdiction as a necessary or desirable supplement to reliance solely on flag states or even to a dual flag and port state regime. The Canadian Arctic Waters Pollution Prevention Act, which authorizes 'pollution prevention officers' to carry out extensive inspections and other regulatory measures,[95] is an instance of coastal state assumption of such jurisdiction to enforce national rules and standards for the protection of that state's own environment and the common environment. The 1972 London Ocean Dumping Convention and the 1973 Pollution from Ships Convention set a further precedent in providing for shared enforcement jurisdiction by flag and coastal states of their multilaterally prescribed rules and standards.[96] (Both these conventions, however, left open for resolution after the Law of the Sea Conference the question of whether or not and how far this competence extends beyond the territorial waters of the coastal state.)[97]

Recognition of the special interests of coastal states in the protection of the marine environment and its living resources is not a radical new concept. As far back as the 1958 Geneva Convention on Fishing and Conservation of the Living Resources of the High Seas, the international community recognized that '[a] coastal State has a special interest in the maintenance of the productivity of the living resources in any area of the high seas adjacent to its territorial sea.'[98] And later, the Statement of Objectives on the Marine Environment located these interests squarely within the framework of a comprehensive management system for ocean resources:

The marine environment and all the living organisms which it supports are of vital importance to humanity and all people have an interest in assuring that this environment is so managed that its quality and resources are not impaired. This applies especially to coastal nations, which have a particular interest in the management of coastal area resources. The capacity of the sea to assimilate wastes and render them harmless and its ability to regenerate natural resources is not unlimited. Proper management is required and measures to prevent and control marine pollution must be regarded as an essential element in this management of the oceans and seas and their natural resources.[99]

What is nevertheless different about the more recent developments is that they not only acknowledge the interests but make provisions more likely to result in their implementation. In the face of an increasingly sharp contrast between coastal interests on the one hand and flag or distant-water interests on the other, there has been some delegation of powers to those states which are most likely to take account of ecological unities. The creation of new alternative enforcement jurisdictions is intended, as will be discussed at length in chapter 5, to see that those states which have particular or exclusive incentives compatible with more inclusive common international interests have the means to pursue them and consequently benefit humanity as a whole.

A major exception that should be noted to this discussion of the application of international environmental law is that the existing international pollution agreements do not apply to certain state-owned or -operated ships, in particular to naval vessels and military support ships. The 1972 London Dumping Convention, for example, stipulates that '[t]his Convention shall not apply to those vessels and aircraft entitled to sovereign immunity under international law,' and the 1973 Pollution from Ships Convention is even more restrictive: 'The present Convention shall not apply to any warship, naval auxiliary or other ship owned or operated by a State and used, for the time being, only on government non-commercial service.'[100] Both conventions nevertheless do provide that the parties shall ensure that such vessels and aircraft act in a manner consistent with their objectives and purposes (with the latter qualifying this by 'so far as is reasonable and practicable').[101] Similarly, the new law of the sea convention will provide that its provisions regarding pollution of the marine environment 'shall not apply to any warship, naval auxiliary, other vessel or aircraft owned or operated by a State and used, for the time being, only on government non-commercial service'; states will at least pledge, however, to ensure that such vessels or aircraft 'act in a manner consistent, so far as is reasonable and practicable, with the present Convention.'[102] Consequently, these exemptions do not necessarily indicate a fundamental conflict between common interests in the prevention of pollution and the narrower national security interests of individual states. In these instances, however, the international community has decided that the particular means of application of environmental policies are inappropriate or insupportable as regards military vessels.

A last word should be added here about the controversy over the effect of differing levels of economic development on the duty to apply and the likelihood of implementation of international environmental law. As mentioned in the last chapter, there has been a considerable amount of controversy at the Law of the Sea Conference and elsewhere on this subject.[103] While differing levels of economic development are irrelevant to the consequences of pollution and arguably should not be taken into consideration in the setting of standards regarding pollution from ships and offshore activities, they are generally recognized to be painfully relevant to the extent to which the policies fixed by the world community can be implemented by particular states at a given time. Without adequate financial and technical assistance, many states may remain incapable of applying some international environmental law.[104]

Termination of ecologically unsound practices and amelioration of the consequences and costs of change

The prescription of new policies does not in and of itself result in the termination of old laws and practices incompatible with them. Consequently, efforts may often have to be made to terminate old crystallizations of opinion and reorientate the decision-making system towards new prevailing patterns of community expectations. As part of this function, special arrangements may be called for to dispose of reasonable and legitimate expectations created under the old regime and to ameliorate some of the costs of change. Problems of *rebus sic stanibus* (changed circumstances) are in no way unique to international environmental policies, but the issue of readjusting selective impacts of the transition to new policies is particularly acute in this field because of conflicting demands on resources between the environmental concerns of advanced industrial states and the development priorities of poorer nations.

Constitutive processes of international organization frequently have to deal with these matters. A good example occurs in the regulation of international drainage basins. The Helsinki Rules specify that each basin state is entitled to a 'reasonable and equitable share' of the beneficial uses of the waters, determined in accordance with certain relevant factors including 'the practicability of compensation to one or more of the co-basin States as a means of adjusting conflicts among uses.'[105] They further provide that one such state may not be denied the present reasonable use of the waters to preserve for a co-basin neighbour a future use of such waters.[106] This combination of principles may often lead to the need for specific adjustment of expectations and practices: 'An existing reasonable use may continue in operation unless the factors justifying its continuation are outweighed by other factors leading to the conclusion that it be modified or terminated so as to accommodate a competing incompatible use.'[107] The commentary on this last article then explains that '[a] modification or termination, to be consistent with equitable utilization, may, in a particular case, require compensation to the user.'[108] The Council of Europe draft fresh water pollution treaty did not deal with these issues of liability and compensation.[109] And a somewhat similar question of what, if anything, to do about possible selective impacts of changes in resource utilization has arisen in the law of the sea negotiations, as a result of fear on the part of some land-based producers of nickel and copper (particularly the former) that future deep sea-bed mining in the international area will adversely affect prices; during the seventh session of UNCLOS III, however, states managed to agree on a formula for production controls for nickel, based on a proposal jointly developed by the United States and Canada.[110]

Compensation will not always, of course, be warranted for those who have enjoyed beneficial use, and it may sometimes be specially called for for those who have not. In the more directly 'environmental' context, governments at Stockholm did recognize the need to terminate or at least modify many ecologically unsound practices for the benefit of all people and their posterity. They further recognized that for the industrialized countries environmental problems are generally related to industrialization, whereas in developing countries most of the environmental problems are caused by underdevelopment itself.[111] It was therefore apparent that most of the burden of change in regard to inclusive resources would have to be borne by industrialized societies, but it was also felt that those states should additionally absorb any costs which might emanate from incorporation by developing countries of environmental safeguards into their own development planning.[112] Consequently, when the UN General Assembly decided that, in order to provide for additional financing for environmental programs, a voluntary fund should be established, it made special provision for the circumstances and requirements of developing countries. The resolution creating the Environment Fund, which was adopted without any opposition, incorporated the additionality principle (that is, that developing countries should receive additional assistance to meet extra costs of development due to present-day environmental concerns), as follows:

[I]n order to ensure that the development priorities of developing countries shall not be adversely affected, adequate measures shall be taken to provide *additional* financial resources on terms compatible with the economic situation of the recipient developing country.[113]

Furthermore, amelioration of the costs of change is not only a matter of financial compensation but may also consist of temporal or other measures designed to achieve orderly and just transition. Such problems arise, for example, in the reconciliation of the preferential fishing rights of coastal states with the traditional rights of others. The *Fisheries Jurisdiction* decision is not incompatible with such a transitional approach.[114] The International Court of Justice held that the Icelandic regulations creating an exclusive fisheries zone of fifty nautical miles were not opposable to the United Kingdom or the Federal Republic of Germany, since these two states have treaty rights and historic interests in regard to the disputed waters and are therefore 'under no obligation to accept the unilateral termination by Iceland.'[115] At the same time, the Court did insist that such legal relations are not immutable:

This is not to say that the preferential and other rights of a coastal State in a special situation are a static concept, in the sense that the degree of a coastal State's preference is to be considered as fixed forever at some given moment.[116]

It therefore concluded that 'the most appropriate method for the solution of the dispute is clearly that of negotiation' with a view towards delimitation of the future rights and interests of the parties.[117] It cannot be concluded from this decision, it follows, that conservation or other needs could never justify limitation or reduction of fishing rights; the finding was rather that, in this case, the particular method of abrupt unilateral termination was unacceptable and unfair.[118] But even if broad extensions of exclusive jurisdiction are the result of general multilateral agreement, that alone does not negate problems of transition. The general point is that the function of avoiding sudden disruptions and dislocations precipitated by rapid change may be necessary, involving a great deal of policy-making with profound economic and social as well as environmental implications.

Appraisal of the ongoing environmental decision-making system

Just as scientific information about the state of the environment is indispensable to the development of effective environmental management principles and practices, so too social scientific information about the functioning of the decision-making system itself is essential for securing their achievement. Various arrangements for appraisal of the manner and measure in which public policies are being put into effect have been built into the international decision-making system. As far as specifically environmental policy is concerned, provisions are being made both for the continuous review of international activities and for periodic or occasional comprehensive reappraisals.

Ongoing appraisal is one of the primary functions entrusted to the United Nations Environment Programme. UN General Assembly Resolution 2997 (XXVII), which created UNEP, listed among the responsibilities of the Executive Director assisted by his secretariat 'to co-ordinate, under the guidance of the Governing Council, environmental programmes within the United Nations system, to keep their implementation under review and to assess their effectiveness.'[119] It further contained the complementary provision that the Governing Council should 'receive and review' the periodic reports of the Executive Director on environmental programs within the UN system.[120] The resolution also set up the Environment Co-ordination Board under the chairmanship of the UNEP Executive Director and under the auspices and within the framework of (and since merged into) the UN Administrative Committee on Co-ordination to bring together the various international bodies concerned in these efforts for collective self-appraisal.[121] Finally, it called upon governments to ensure that appropriate national institutions 'shall be entrusted with the task of the co-ordination of environmental action, both national and international.'[122]

Subsequently, the UNEP Governing Council has taken a broad view of the appraisal function and UNEP's mandate as regards appraisal and co-ordination. It requested the Executive Director 'to establish ... an inter-agency task force which would devise a methodical way of providing data concerning system-wide activities related to the environment for the purpose of recurrent review' and additionally 'to formulate guidelines for the content of national reports on current environmental activities.'[123] The UNEP secretariat has, in fact, already begun this recurrent Review of the Environmental Situation and of Activities Relating to the Environment Programme.[124] The idea is that the recurrent review should be prepared not only to serve the purpose of identifying environmental problems and defining management objectives but also to constitute an extended management information system and a means for identifying gaps in collective action. Among other things, it attempts to summarize the overall expenditure in dollars and man-years of the UN family devoted to environmental activities – proportionally, relatively, regionally, by subject area, and otherwise.

While UNEP has an overall mandate with regard to evaluation and co-ordination of international environmental activities, many other organizations perform the appraisal function with respect to particular sectoral or regional dimensions of the 'environmental' problem. The Intergovernmental Maritime Consultative Organization (IMCO), for instance, in establishing its Marine Environment Protection Committee (MEPC), charged the new body not only with executory functions under IMCO conventions but also with the responsibility 'to co-ordinate and administer ... the activities of the Organization concerning the prevention and control of marine pollution from ships.'[125] And to take the North Atlantic region as an example, several existing European organizations (for example, the Council of Europe, the UN Economic Commission for Europe, the European Communities, and the Council for Mutual Economic Assistance) have recently spawned environmental 'watchdog' committees,[126] and the regional Oslo Dumping Convention and the Paris Land-Based Sources Convention each set up a special commission whose primary duties centre on the supervision and review of control measures taken in their implementation;[127] in fact, even the North Atlantic Treaty Organization (NATO) in 1969 set up a Committee on the Challenges of Modern Society (CCMS) to 'examine how to improve, in every practical way, the exchange of views and experience among the Allied countries in the task of creating a better environment for their societies ... ' in addition to itself launching 'pilot projects' towards the same goal.[128]

Although continuous evaluation is essential, it has often been found desirable to have an occasional watershed reappraisal. The results may be unsettling and at the same time beneficial. The most comprehensive recent enterprises of this sort are the report of the Pearson Commission (which appraised the activities of the International Bank for Reconstruction and Development), the Jackson 'Capacity Study' (which evaluated the United Nations development system), and the recent Ad Hoc Committee on Restructuring (which took on the whole

UN economic and social sectors).[129] The most colourful conclusion of these reviews is probably the eclectic conclusion of the Jackson report that the decision-making and administration structure was 'incapable of intelligently controlling itself': 'In other words, the machine as a whole has become unmanageable in the strictest use of the word. As a result, it is becoming slower and more unwieldly, like some prehistoric monster.'[130] In more conventional diplomatic parlance, however, the Restructuring group reached similar troubled conclusions, and the General Assembly is already trying to take measures to implement some of its recommendations for organizational change.[131]

The United Nations Environment Programme was founded on a different philosophy and without the comprehensive operational ambitions of the UN Development Programme and many other agencies, at least partly in the hope of avoiding having it, too, repeat the history of the dinosaur. But this has not been taken to imply that UNEP and the international environmental decision-making system need not be watched closely and perhaps entirely overhauled from time to time. At Stockholm, governments adopted a resolution recommending convening of a second UN Conference on the Human Environment,[132] and the UN General Assembly took note of the proposal at its next session.[133] The matter was referred to the UNEP Governing Council, which a year later and again a year after that did recommend a second Human Environment Conference (not before 1980).[134] There is, of course, a great deal of scepticism about the basic utility of such general multilateral conferences, but they are at least acknowledged to provide a benchmark or occasion for reappraisal, if not a means thereof.

International environmental funding and some related matters

Funding, strictly speaking, is not a separate function of a decision-making system, but rather a critical underlying condition for the performance of all the functions thereof. Finding or raising funds is nevertheless a question which commands a great deal of attention from the international community as a whole and which consumes a great deal of time on the part of most international organizations. In the environmental context, two basic types of financial support problems arise: financing to support projects whose primary objective is environmental restoration or preventive action, and supplementary funding to meet the additional costs of development and other ongoing activities to the extent required for building in environmental protection measures (including both the additional costs which are desirable from the point of view of inclusive international environmental interests and the extra costs a country or other actor might be encouraged to incur solely on the basis of its own long-term interests). A third type of budgetary issue is presented by the administrative costs of maintaining the infrastructure of the United Nations Environment Programme and other international environmental bodies. Various arrangements are being made by the international system to meet these costs by means of voluntary funds and through UN budgetary assessments. Certain indirect measures, such as prohibition of the use of countervailing duties and other discriminatory practices in connection with environmental policies, are also relevant to questions of who pays and how to finance international environmental co-operation.

Reference has already been made to the UN Environment Fund, which is a voluntary fund established by the General Assembly following a recommendation of the Stockholm Conference. Its purpose is to 'finance wholly or partly the costs of new environmental initiatives undertaken within the United Nations system.'[135] Accordingly, it is to finance 'programmes of general interest' such as

regional and global monitoring, assessment and data-collecting systems, including, as appropriate, costs for national counterparts; the improvement of environmental quality management; environmental research, information exchange and dissemination; public education and training; assistance for national, regional and global environmental institutions; the promotion of environmental research and studies for the development of industrial and other technologies best suited to a policy of economic growth compatible with adequate environmental safeguards: and such other programmes as the Governing Council may decide upon ...[136]

In addition, as has already been discussed, it is to provide extra funds to developing countries to enable them to pursue their development priorities in an environmentally sound manner.[137] The Environment Fund does not, however, meet the costs of servicing the UNEP Governing Council and secretariat, which are borne by the regular budget of the United Nations.[138]

At the Stockholm Conference, the Environment Fund was projected to be $100 million over the first five years. The United States pledged to provide up to 40 per cent, on a matching basis, over the initial period. Japan promised up to 10 per cent additional if major developed countries made substantial contributions, and Sweden, Canada, Australia, and the Netherlands also offered specific pledges. Several other expressions of general support were forthcoming, but without a price tag.[139] After Stockholm, the initial period of the Fund's activities found voluntary contributions coming in only at a slow rate, and the lack of available funds made it impossible for many projects to be approved at their scheduled time; by the second annual session of the Governing Council, however, things had improved, and resources were at a higher level.[140] Overall, the total paid and pledged for 1973–8 has been somewhat above anticipated levels, and the allocation for Fund Programme activities is now running about $30 million annually.[141] In addition, of course, the UN Fund for Population Activities and the UN Habitat and Human Settlements Foundation must be regarded as sources of international financing for environmental or environment-related activities. Their total is, of course, pitifully small compared to the many billions of dollars spent each year for defence and security purposes, but they represent a relatively impressive amount in the context of regular UN budget appropriations.

Besides direct funding questions, there are various possible indirect measures for meeting or compensating for cost differentials in international trade and development. The Organization for Economic Co-operation and Development has, however, determined their use to be unnecessary and undesirable as part of an international environmental management system. The OECD has instead adopted as a guiding concept the 'polluter pays principle' of cost allocation, whereby the costs of pollution prevention and control measures are to be reflected in the costs of goods and services which cause pollution in their production or consumption.[142] Its 1972 Guiding Principles on the Environment acknowledge that different national environmental policies are justified by a variety of factors, including, among other things, different pollution assimilative capacities of the environment in its present state, different social objectives and priorities attached to environmental protection, and different degrees of industrialization and population density.[143] But they specify that the policy of the Organization is definitely that, in conformity with the provisions of the General Agreement on Tariffs and Trade (GATT), any measures taken within an environmental policy framework regarding polluting products should be applied in accordance with the 'principle of national treatment' (that is, identical treatment for imported products and similar domestic products) and with the 'principle of non-discrimination' (identical treatment for imported products regardless of their national origin).[144] As a result, compensatory trade and export measures to adjust for environmental protection costs are deemed unacceptable: 'In accordance with the provisions of the GATT, differences in environmental policies should not lead to the introduction of compensating import levies or export rebates, or measures having an equivalent effect, designed to offset the consequences of these differences on prices.'[145] Two years later, the Council of the OECD adopted a Recommendation on the Implementation of the Polluter-Pays Principle, which states that it constitutes for member countries 'a fundamental principle for allocating costs of pollution prevention and control measures introduced by public authorities' and makes certain specific recommendations for its uniform observance.[146] Meanwhile, the Council of the European Communities has taken its own initial step in adopting a Recommendation on the Application of the Polluter-Pays Principle, which accepts similar standards for cost allocations and actions by public authorities in the field of the environment in its member states.[147]

Finally, although the international community has decided to accord great solicitation to the development priorities of developing countries, their quest for higher standards of living does often entail deliberate or inadvertent modification of the natural environment to achieve economic objectives. Where there are adverse ecological consequences, such activities do involve increased costs to society, which ought to be measured. To the extent that they are not taken into account, the difference often has to be paid at a later stage of development. Since the higher costs of future remedial action may frequently be avoided by prudent planning and early preventive measures, the International Bank for Reconstruction and Development (IBRD or World Bank) has been examining criteria for the identification and analysis of environmental considerations to be used in evaluating projects and considering requests for its development assistance loans.[148] The United States Agency for International Development (AID) has also initiated similar considerations.[149] So, while these agencies have, of course, been intent on their development funding, they too have made an effort to internalize environmental costs to the activities which generate them in order to reduce overall societal costs in the long run.

PART THREE

**Analysis of Trends in Environmental Order
and Recommendation of Future Policies**

5

Towards an international
ecological law and organization

Analysis of the trends described in the preceding two chapters shows that neither the doomsday prophets nor the world federalist cheermongers have an accurate understanding of ecological unities and their incorporation in the present world public order. It furthermore tends to discredit the excessive pessimism of those who claim that nothing significant can reasonably be expected to be accomplished given the selfish drives of nation-states. Rather, the overall course of events seems to support the highly qualified optimism of others who feel that, given adequate information and comprehension of the scientific and political unities, international politics can evolve towards a more ecologically sound order. The reordering process may already be well begun, but it has a long way to go under very intense time pressures.

Achievements to date are inadequate, and the international approach has been considerably lacking in theoretical and practical comprehensiveness and consistency. World community expectations remain unrealistic in many ways, or, as one prominent analyst has summarized the situation, 'existing international environmental law is inadequate, both in scope and substance – in scope, in that it is incomplete, and in substance, in that it is inconsistent, fragmentary and in large part inchoate.'[1] The record of putting sound environmental policies into practice is more unimpressive still, and future constitutive arrangements remain unclear in such fundamental areas as jurisdiction to set and enforce regulations, criteria and methods for assessing liability and ensuring compensation for environmental damage, dispute settlement principles and mechanisms, and so on. Thus, although the evidence shows that certain of the conditions called for by the logic of collective action for satisfying common environmental interests are being created, the trends are in some measure conflicting and incompatible, and the whole approach lacks a coherent conceptual framework for dealing with the environmental dimension in the context of the total world public order.

'Custodianship' and 'delegation of powers'
as the conceptual framework

As was discussed in chapter 2, the most general goal of the world public order is that of furthering common and rejecting special interests. The basic common interest of the world community in the minimization of injury to the environment and its constructive use for present and future generations has both inclusive or shared and exclusive or non-shared aspects, which are somewhat interchangeable in character when viewed from different perspectives or with differing expectations. In order to promote fulfilment of common environmental interests, three basic strategies are available. It may often be advisable to internalize the environmental externalities to one specific actor by recognizing or creating an exclusive interest so that there will be an incentive to take them into account in deciding on policies (providing, of course, exclusion is itself possible and not too costly). In other instances where exclusion is not viable or desirable, it may still be feasible to attribute environmental costs to specific functional activities, so that actors will have to take them into account in deciding how much of an activity to engage in (and what compensation is due to others as a result). Finally, there are some cases, particularly with necessarily inclusive interests in shared resources such as the biosphere or ocean waters, where collective action by the international system is greatly preferable or even essential in order to control certain acts or activities so as to prevent injury and foster the protection of the environment in the light of the multiple values of the world community (that is, not only wealth and power, but all the values of human dignity). Large-scale collective action requires a degree of compulsion, and, in the absence of central policing capabilities, reliance will have to be placed to a great extent on states whose own self-interests motivate them voluntarily to exert the effort. The basic question is, keeping common environmental and other goals in mind, how to combine these diverse elements into a rational environmental management system within the overall world public order.

In the law of the sea negotiations, Canada has advanced the concepts of 'custodianship' and 'delegation of powers' as ways of achieving common interests through reliance on individual motivations. More specifically, the policy – as set forth in several statements and articles by J. Alan Beesley, Canadian Ambassador to the Law of the Sea Conference, and others – provides that certain particular interests of coastal states in adjacent areas should be recognized but understood to be accompanied by concomitant duties to the international community as a whole.[2] It has been summarily promulgated in the third 'Ottawa Principle' on the rights and duties of coastal states:

The basis on which a state should exercise rights or powers, in addition to its sovereign rights or powers, pursuant to its special authority in areas adjacent to its territorial waters, is that such rights or powers should be deemed to be delegated to that state by the world community on behalf of humanity as a whole.[3]

The essence of the interrelated concepts of custodianship and delegation of powers has elsewhere been summarized in terms of three fundamental attributes:

first, the primary or priority interests of the coastal state in all activities in areas of the sea adjacent to its shores must be reflected in international law; second, much of the administration of the law of the future must be 'delegated' to the coastal state and must be based on resource management and environmental management concepts; third, the basis for an accommodation between conflicting interests in the uses of the sea must lie in a better balance between the rights and consequent responsibilities of states, and hence the coastal state must exercise both its existing sovereign powers and its future 'delegated' powers not only in its own interests but as 'custodian' of vital community interests in the uses of the sea, on the basis of internationally agreed principles to this end.[4]

This Canadian policy and the twin concepts in which it has thus been encapsulated have been applied to a whole range of international law of the sea issues and they are said to be fundamental to domestic maritime and fisheries legislation as well.[5]

The concept of custodianship in conjunction with delegation of powers is generalizable to other areas of environmental public order. (Analogous to the interests of a coastal state in guarding against undesirable impacts on its own interests by pollution of adjacent waters, for example, a state has a readily recognizable interest in self-protection against the effects of weather and climate modification activities in adjacent areas which will be to its disadvantage.) Although it has not been explained in precisely these terms, what the policy does in effect is expressly acknowledge that it recognizes exclusive interests within the overall common interest rather than new individual sovereign or special national interests *per se*. It takes account of the logic of collective action in its basic premise that rational, self-interested individuals or states will not voluntarily act to achieve their common or group interests but will seek to further their own exclusive or national self-interests. Therefore, if there is legal recognition of interests and consequent delegation of powers to those states whose own interests coincide or are compatible with wider inclusive common interests in a given dimension, the incentive to act will be coupled with the right to do so – that is, authority and control are cojoined – and the desired outcomes may be achieved by the world public order.

Within this general conceptual framework, it is now possible to appraise the trends in international environmental law and organization and to make recommendations for the future in both areas; some observations can also be made about the inherent defects in the framework itself as they reflect both the lack of 'pure' cases of the theory and the peculiarities endemic to the system to which it is applied. Still, as a general theoretical guideline, it remains the most promising approach for present purposes.

International environmental law to date

International law, defined as the expectations and practices of the world community, obviously has a high political content. It is, of course, not totally a pragmatic justification of political and other exigencies, as it embodies expectations about what ought to be as well as what are the present implications of relationships among nations and other actors. No problem in international law, however, can be viewed realistically without its context of underlying political, economic, sociological, scientific, technical, and other factors, and no viable solution to an international law problem can be achieved which does not accommodate these contextual realities. The interpretation of the law is thus a two-pronged and interconnected process: one must review the manifest content of the relevant texts or different legal expressions according to harmony or contradictions and in terms of concrete implications of abstract doctrines; and one must also reconstruct the expressions in the light of the actual situation and the shared subjectivities of the actors as they appear to an outsider.[6]

I shall accordingly attempt to appraise both the strengths and weaknesses of international environmental law to date and its efficacy in response to the emerging ecological context. I shall additionally assess the trends in the light of the concern of the world community for the maintenance of the world public order itself.

To repeat an earlier observation here, many of the expectations of the world community are changing so as to incorporate new ecological conditioning factors. Two strategies of 'general' or 'market' deterrence were identified in chapter 2 as efficacious means of dealing with environmental externalities: first, where feasible and not too costly, it may be desirable to 'internalize' the calculations of costs and benefits of a functional activity to a single actor; and second, while such exclusions may not be possible or practical in certain other cases, it may still be viable for the actor that generates an environmental effect to come to an agreement with the recipient (s) on the proper level of the effect – perhaps with reparations, compensation, or damages of some kind from one to the other(s). The trends described in chapter 3 towards determination of competence over resources clearly reflect movement in the former direction, and many of the trends in the regulation of resource use can be seen to represent resort to the latter alternative. Other regulatory developments help to establish the basis for the third available strategy – direct control by the international community of certain activities with highly significant environmental consequences – which are discussed below (see 'Evolution of international environmental organization,' 124–35).

COMPETENCE OVER RESOURCES

Inclusive and exclusive interests require accommodation at every step, both between and within the two broad categories. From the point of view of the greater shaping and wider sharing of human values, it is desirable to maintain as high a degree of inclusive competence as possible. Any new exclusive claims should consequently be justified only in terms of custodianship on behalf of the wider community, with powers delegated expressly on that condition – in other words as measures of expediency in the light of limitations inherent in the contemporary international decision-making system rather than as extensions of outmoded sovereign prerogatives. The basic challenge of accommodation is, it follows, while giving primacy to inclusive interests, to ensure the protection of exclusive interests without giving inadvertent protection to interests which are special and hence incompatible with common interests.

What this approach implies for both time-honoured exclusive rights and newly acquired exclusive competences in terms of custodianship and delegation of powers has been succinctly summarized by Leonard Legault:

Where acquired well-established rights are concerned, as in the case of the coastal state's sovereignty over the territorial sea or its exclusive sovereign rights over the continental shelf, the notion of custodianship implies that the coastal state must exercise those rights with due regard to the shared interest of all states in, for instance, innocent passage through the territorial sea and freedom of navigation in the superjacent waters of the continental shelf. From this point of view the notion of custodianship has a self-denying effect; it highlights the limitations already inherent in various recognized forms of maritime sovereignty and jurisdiction under traditional law, and presents them as positive duties owed by the coastal state to the international community. Where new or extended rights are sought to be acquired, such as anti-pollution authority in areas adjacent to the territorial sea, custodianship rides piggyback on the concept of delegation of powers (which, of course, does not apply to sovereign powers already acquired). Thus, on the one hand the concept of delegation of powers is acquisitive in effect and serves as a legal fiction (in the best sense of that term) under which legitimate aspirations of the coastal state can be satisfied; on the other hand it also carries with it the self-denying aspects of custodianship and hence does not open the door to unbridled arbitrary action by the coastal state. The idea, of course, is that states should not claim benefits without accepting corresponding obligations. (The concepts of custodianship and delegation of powers can, of course, be applied even to the right of flag states to navigate the high seas, which should also entail certain duties and responsibilities.) That some states may claim the obligations in order to gain access to the benefits in no way detracts from the essential objective of balancing rights and responsibilities.[7]

(The above discussion was solely in terms of law of the sea issues but is more broadly applicable to areas such as weather and climate control, air pollution, acquisition and use of resources in space, Antarctica, and so on.)

Inclusive resources
Turning now more specifically to the observed trends in international environmental law, there has been a strong tendency to carve out of formerly inclusive resources new exclusive competences on a functional basis. As has been said, this can be seen to have the effect of internalizing the environmental externalities of the activities in question and so is theoretically defensible, provided the costs of such exclusion are not too great in terms of other common interests of the world community. As far as fish and resources of the continental shelf are concerned, international law has clearly come to recognize two-hundred-mile exclusive economic zones and shelf jurisdiction out to the outer edge of the continental margin. By contrast, exclusive competence for decision-making on pollution control standards in zones of similar compass is still most often considered too costly in interfering with common interests in traditional freedoms of the seas (for example, navigation, transportation, commercial activity, and communications uses), especially if this competence is defined to include standard-setting in regard to ship design, construction, manning, and equipment, as well as in respect of dumping, operational discharge, and traffic routing and separation. Exceptions to international uniform pollution standards have nevertheless already been allowed for certain internationally designated 'special areas', and recognition of the residual competence of a coastal state also seems warranted where particularly severe climatic or hazardous ecological conditions raise the possibility of major harm or irreversible damage to the environment (including living resources). As regards other functional resource considerations, such as those concerning

anadromous species of fish, the costs and benefits of internalizing competence have not yet been fully balanced, but a shift to exclusive control by the state of origin and return seems virtually certain in any new law of the sea 'umbrella treaty'.

Where inclusive competence is retained – because weighing of common community interests, characteristics of the resources themselves, equitable considerations, or some other factors so dictate – there is no substitute for collective action (specific parameters of which will be discussed later). There will have to be not only a comprehensive system for conservation and management of the living resources of inclusive areas, but also equally expansive arrangements for the exploration and exploitation of non-living resources, both within the overall context of ensuring protection of the total environment. Since voluntary action and contributions in support of common interests are so difficult to engender, it is highly advisable that substantial portions of the future revenues to be derived from resources under the competence of the new International Sea-bed Authority be retained for the central community. They could then be available for side-payments to induce co-operation and equitable sharing and for the setting-up of various regulatory mechanisms to oversee compliance (and perhaps also for the general support of international organizations).[8] Should resources from outer space become available on a commercially feasible basis, similar considerations would apply.

Exclusive resources
As far as long-established exclusive resources are concerned, the recognition in Principle 21 of the Stockholm Declaration that sovereign rights imply responsibilities to take account of the environmental interests of other states and of the common environment[9] is becoming deeply embedded in international environmental law, if indeed it has not always been inherent in traditional notions of international law. It is, of course, essential to the concept of custodianship. Much energy has been consumed in debating whether this is a limitation placing the traditional concept of sovereignty in its environmental context or whether, on the contrary, it is at best hortatory, since sovereignty cannot be subject to qualification or limitation. Such legalistic hair-splitting is both theoretically and practically dysfunctional. Modern technological factors have increased conditions of interrelatedness, thereby causing a change in the whole context of policy-making. This critical change of condition has inexorably influenced the law, and Principle 21 and similar statements can be understood as merely registering cognition of the fact.

There are certain functional areas, for example weather and climatic effects, where notions of both inclusive and exclusive competence should be expanded. As activities which inadvertently or deliberately have such effects on a large scale become more prevalent, so do the dangers of harm to inextricably shared interests, and there is a need for an overall international regime to deal with these problems. Recommendation 70 of the Stockholm Action Plan,[10] although very narrow in its conceptions, did at least acknowledge the issue; and the Convention on Prohibition of Military or Any Other Hostile Use of Environmental Modification Techniques[11] is commendable for its recognition of the necessarily inclusive features and is a start – but only a start – in the right direction. And as far as the exclusive aspects are concerned, it is probably axiomatic from traditional concepts of impact territoriality that a state's sovereign land and air rights necessarily include competence over the weather and climate of its territory. Thus, while the new Agreement Between the United States and Canada relating to the Exchange of Information on Weather Modification Activities is admirable in '[r]ecognizing the desirability of the development of international law relating to weather modification activities having transboundary effects,' it seems excessively timid in its qualification that nothing therein is to affect the question of responsibility or liability in such cases 'or to imply the existence of any generally applicable rule of international law.'[12] In the light of emerging ecological interdependencies, it must be expressly understood to follow also that states have the responsibility to ensure that activities within their jurisdiction or control do not cause significant weather alterations in other states or detriment to common atmospheric or climatic conditions.

Obviously questions of competence are inherently complex, and they are, of course, often also inseverable from questions of proper resource use. At times, conflicting common interests may seem to produce an equilibrium, or practical obstacles may seem to be insurmountable, and then a pragmatic search for *an* acceptable solution is necessary. Pollution control in international straits may be a case in point. Certain governments are especially adamant in asserting that common security interests require the preservation of maximum freedom of transit for military and support vessels and aircraft through and over these corridors; other states, however, point out that the environmental risks of navigation through these passages are also of exceptionally serious concern and that coastal states should therefore have the right to set and enforce higher standards for their own protection and for common environmental protection. In the face of such strong and deeply held directly conflicting convictions, any resolution of the problem would have to represent an exercise in circumvention or a tentative 'fudging' of potentially irreconcilable interests. The present provisions drafted by UNCLOS III provide coastal states with the minimal authority to make laws and regulations for the prevention, reduction, and control of marine pollution 'giving effect to applicable international regulations regarding the discharge of oil, oily wastes and other noxious substances in the strait'; even these laws and regulations, however, it is expressly stipulated, 'shall not discriminate in form or in fact amongst foreign ships or in their application have the practical effect of denying, hampering or impairing the right of transit passage.'[13] About the best that can be said for these provisions from the environmental point of view is that they are at least a start, and that they represent an exercise by diplomats in the 'art of the possible.'

In sum, the predominant trend in determining competence over resources is towards increasing exclusive competence on a functional basis. This has the effect of internalizing the problem of deciding on the best adjustments to be made in the light of all the attendant costs and benefits – in particular of environmental considerations – to one decision-making unit, that is, a single nation-state. Since the costs are by nature social or sociological as well as economic or monetizable, this approach has the advantage of allowing for a certain degree of subjective analysis by observers close to the particular difficulties involved, as opposed to long-distance eclectic assessments. The conservation and management of fishery resources at some optimum sustainable yield, for example, involve not only direct ecological interrelationships (food chains and maximum biologically sustainable yields), but also economic conditioning factors (the costs of obtaining alternative yields in the same areas) and social considerations (such as the dependency of coastal populations on fishing for a livelihood). These factors are more likely to be successfully accounted for and accommodated on a localized basis than as part of aggregate regulation from afar. The wholesale extension of exclusive competence does, nevertheless, carry with it the profound danger that overall common interests will be lost from sight in the pursuit of private gain. A coastal state may possibly, for instance, decide to overfish or license overfishing of its stocks now in the hopes of rectifying a pressing trade or balance of payments problem or to obtain immediate investment capital at the expense of future yields. It is arbitrary and irresponsible action along such lines that the concepts of custodianship and delegation of powers, with their essential implication of international overview and residual powers, are designed to forestall or minimize.

REGULATION OF RESOURCE USE
Many trends in the regulation of resource use, to reiterate, can be seen as representing attempts to calculate the environmental costs and benefits of

certain specific activities and to attribute responsibility and/or liability for their control and any ensuing external costs to a particular actor. Without this initial legal delimitation of rights, there would be no basis for bargaining, and 'general' or 'market' deterrence could not be relied upon to transfer and re-combine the factors so as to maximize the production of all values at stake. In a world free of transaction costs, the same allocation of resources would come about under any initial assignment of liability between all the parties involved; that is, bargaining would take place until the final result maximized the value of production.[14] In the real world, however, transactions do have costs, and so do substitutes for transactions (for example, direct central regulation of activities), and there is therefore a need to determine the best cost-avoider; it thus matters greatly both how liability is placed in the first instance and what are the relative scope and costs of alternatives to market exchange. At the present stage, information about social and other costs is very imprecise, but workable solutions to pressing problems cannot wait until all the facts are in. It is consequently largely necessary to proceed on a case-by-case basis as particular environmental threats become apparent; but it is to be hoped that the theory and observations of this study will help in suggesting ways in which responses can be developed and implemented on a coherent, systematic basis as part of a comprehensive approach to environmental protection.

Controlling injurious use

As far as controlling injurious use of resources is concerned, the most obvious recommendation is to repeat the exhortation of Principle 22 of the Stockholm Declaration that states should 'co-operate to develop further the international law regarding liability and compensation for the victims of pollution and other environmental damage ...'[15] While the right to compensation for pollution damage undoubtedly already exists both in many treaties and under customary international law, difficult questions remain unresolved in relation to satisfaction of the right, especially with regard to compensation for damage suffered beyond the limits of national jurisdiction. A variety of modes could be devised for ensuring compensation of victims of environmental damage – ranging from international funds or insurance schemes to private rights of action under the laws of each state in accordance with internationally agreed obligations and, in appropriate circumstances, to direct compensation by the responsible state. But the establishment of the fundamental underlying principle of state responsibility for environmental injury is itself crucial if states are to be motivated to develop procedures for the assessment of damage, the determination of liability, the payment of compensation, and the settlement of related disputes.

In relation to inclusive resources, it is axiomatic that international legal regulations must keep pace with technological advances and with political changes. It follows that the new law of the sea, for instance, must take full account of the environmental implications of such modern innovations as nuclear ships, supertankers, deep-water ports, bulk ore carriers, icebreakers, sea-bed mining technologies, submarine scanning devices, and so on. It has also, of course, to keep up with concomitant political and economic developments, for example, the activities of multinational corporations in the exploitation and transportation of resources, the formation of international consortia for sea-bed mining and development, changing military postures, and so forth. As far as the atmosphere is concerned, again, Recommendation 70 at Stockholm suggests that '[g]overnments be mindful of activities in which there is an appreciable risk of effects on climate,'[16] and it follows that there should at least be rapid recognition by governments other than those of Canada and the United States of their duty to consult and to avoid activities posing a serious threat along these lines. Broadly understood, such threats may range in scope from large-scale weather modification programs to the effects of supersonic flight on the atmosphere and to damage to the ozone layer as a result of the use of aerosol sprays. The time is ripe for neighbouring countries to conclude bilateral and regional conventions on local weather modification activities and to begin negotiations leading to multilateral regulation of activities that affect or may affect the planetary weather and climate (beyond the prohibition of military uses in the new environmental modification treaty). And in the cases of space, international rivers, Antarctica, and other currently inclusive resources, as has been indicated, states cannot postpone indefinitely or even for very long working out such problems as the implications of general international law doctrines of state responsibility and liability as they apply under new circumstances and to the particular legal needs of each resource.

Many of these comments apply equally to those resources under exclusive competence. Principle 21 of the Stockholm Declaration proclaimed that states have 'the responsibility to ensure that activities within their jurisdiction or control do not cause damage to the environment of other States or of areas beyond the limits of national jurisdiction.'[17] Principle 22, quoted above, was a necessary consequence of Principle 21, since the international law of state responsibility and liability in general, and particularly in the area of environmental injury, has been so little developed. Another logical outgrowth of this responsibility is the duty of states to inform one another concerning the possible environmental consequences of their actions upon shared resources and/or upon areas beyond their jurisdiction, and it is regrettable that – even with General Assembly resolution 3129 (XXVIII), the OECD Principles Concerning Transfrontier Pollution, and the notification and environmental assessment articles drafted at UNCLOS III[18] – the duty of notification has been neither adequately defined nor fully recognized at international law.[19]Some encouragement can be derived from the fact that the Executive Director of UNEP has been given a mandate and has commenced work on the development and application of these and related environmental law principles;[20] but clearly it is not an easy task in the face of the inertia, apathy, antagonism, and opposition even to the basic principles already enunciated and more broadly to the progressive development of internationl environmental law.

In addition, the substantive dimensions of state environmental responsibility should be reviewed and interpreted more realistically and expansively to encompass wider areas of activities with significant environmental effects, and a lot of implications and recommendations will derive from this. For example, the regional Convention on the Prevention of Marine Pollution from Land-Based Sources[21] should be extended to cover the oceans as a whole and eventually to deal with pollution that reaches the oceans by way of the atmosphere. Finally, in the special case of newly acquired exclusive rights and responsibilities, as has already been discussed, any powers so delegated should be defined to be conditioned upon their exercise only in a manner consistent with common interests, that is, on the basis of custodianship.

At the very minimum, it follows, the international community must undertake to delimit and oversee its own interests, whatever degree of collective action may be attempted to protect and advance them. The basic recommendations above having to do with the development and specification of the concept of state responsibility require a certain amount of international co-operation in lawmaking but may then be largely implemented through private or self-interested actions of nation-states in seeking their legal rights. Other recommendations – especially those concerned with the establishment of viable regimes for the high seas, the atmosphere, planetary weather and climate, and other necessarily inclusive resources – do require continued exertion of some degree of effective control by the international system or elements thereof; and on pages 124–35 below ways are suggested in which the contemporary public order can be better organized to supply this in the absence of central coercive institutions.

Facilitating constructive use
In the meantime, however, something should be said about the other aspect of the regulation of resource use: facilitating constructive use of resources. Similar basic theoretical and practical considerations apply to facilitating beneficial side-effects or positive externalities through conservation programs and other management measures as were pertinent to controlling pollution and

other negative environmental externalities. In certain cases it may be feasible to internalize the activities in question so that a single actor will have an incentive to promote conservation and rational management, and in others it may be possible to effect agreements among a small number of interested parties; but in some resource areas there must be broad-based regulatory regimes. Exclusive two-hundred-mile fisheries or economic zones and other such competences have already been discussed, so the focus here is on the other two strategies.

As far as inclusive resources are concerned, a matter of highest priority from the ecological perspective is the organization of a viable regime for the conservation and management of living marine resources. The need for international agreement and action towards this end was recognized as far back as the 1958 Geneva Convention on Fishing and Conservation of the Living Resources of the High Seas,[22] but it is only recently that the depletion of stocks and resource scarcities have impelled states to try to clarify and give effect in a workable manner to the interests recognized at that time. As far as implementation is concerned, it is clear that a great deal of reliance will have to be placed on regional fisheries commissions, on coastal states, and on states with historical preferences with regard to certain stocks, and more will be said on these matters subsequently. It has already been noted that anadromous species are truly a special case, in that most effective conservation and management measures can only be taken in the rivers and streams of their state of origin and usually involve considerable costs to that state (not only of a directly monetary nature, but also in terms of forgoing hydroelectric utilization and other interdependent factors). They are not, however, the only special case – and what have states individually and collectively managed to do lately for whales and other cetaceans (marine mammals)?

As to exclusive resources, adequate information is as relevant to enabling states to facilitate constructive and compatible resource use as to providing warning of negative environmental impacts. In addition, in the case of newly acquired exclusive zones the concept of custodianship comes back again in full force. In order to internalize cost-benefit accounting, a coastal state should have clear authority to regulate and control the exploitation of coastal species; but, in order to accommodate remaining commonalities of interest, this power should be delegated on the basis of internationally agreed principles and subject to review of its exercise by an international tribunal in the event of disputes with other states.

These considerations do not apply to non-renewable or 'stock' resources in quite the same manner as they do to the renewable or 'flow' resources[23] in coastal areas. There are fewer externalities involved in the exploitation of the former, and they are insufficient to provide a conservation or environmental justification for their relegation to exclusive competence; in addition, or consequently, use of these resources does not require the same type of constant regulation and active management. By contrast, there are strong countervailing considerations – including the opportunity to preempt a new source of revenue for the central community and its institutions and certain equitable arguments – militating in favour of retaining such resources under inclusive competence. It has nevertheless become obvious that coastal states will be permitted to maintain claims to the resources of the continental shelf out to two hundred miles and beyond that through the natural prolongation of the continental margin (with some revenue sharing as to resources between the two-hundred-mile mark and the outer edge). Accepting this seeming inevitability, the recommendation here then becomes for the international community to devote renewed energy to the development of and co-operation with a viable regime and international authority for the exploitation of the resources of the deep sea-bed. The aim is not only to avoid as far as possible future acquiescence in territorial expansionist claims, but also to establish in real and concrete form the principle of the 'common heritage of mankind.' Similarly, it is past time to begin thinking of a regulatory regime for the resources of outer space, in anticipation of their commercial significance at some future date. It should be added, however, that in so far as there are some undesirable externalities associated with the exploitation of these non-renewable resources, in particular the possibility of pollution from sea-bed-based operations, they should be dealt with by the appropriate competence responsible for their management. There must be at least international minimum standards for the protection of inclusive interests, with the possibility of more stringent rules and standards being promulgated by coastal states for their zones and by flag states for their ships anywhere.[24]

Regulation of the use of resources, both in its negative and positive aspects, is an exceedingly complex polemic and cannot be considered entirely distinct from the question of determination of competence over resources. I have made certain basic recommendations on this matter above, which should help to achieve a more environmentally sound legal order henceforth. In addition, the specific inadequacies of the abundance of treaties and other international agreements, caselaw precedents, and varied examples of behaviour were discussed in chapter 3. It goes without saying that, in the course of the elaboration, interpretation, and revision of this legal order, the international community would be well advised to rectify these deficiencies. Here I shall rest content with a few more observations on the third topic discussed in connection with resource regulation: the planning and development of new uses of resources.

Planning and development of new uses
As concerns inclusive competences, there is a most urgent need for an overall method of integrating information inputs, planning, and promotional activities with respect to all the components of the earth-space environment. The Governing Council of the United Nations Envi;onment Programme is, of course, to some extent already charged with these functions, but it has not been endowed with the capabilities for their continuous performance. Aided by such broadly inclusive planning, actual development should go forward on a regional basis (as most ecological problems, though globally interdependent, tend to have a regional focus or centre of gravity). This would make use of the small group effect identified in chapter 2 as critical to securing action in the common interest[25] – that is, in certain small groups, both the benefits of group co-operation to each individual member and the contributions of each of them to the objective being sought are most readily apparent. Such regionalism must, of course, take account of the larger ecological unities and be organized in accordance with natural ecological subsystems rather than synthetic political agglomerations alone.

In regard to exclusive resources, comprehensive planning and development are needed in all countries at all levels – from local to metropolitan, to drainage basins, to larger and ultimately national and transnational levels. Many countries (and all the major industrial countries) already have environmental agencies or ministries responsible for overseeing and managing their own environment and its resources. Some such environmental planning has reached an impressive stage of sophistication.[26] Co-operative planning for exclusive resources – as was illustrated by the United Nations Conference on Human Settlements – may often be useful and economical, but it is also essential that responsibility be delegated to and assumed by continuing organizations which can implement the policies by actually carrying out the planned development. In short, if a nation does not look after its own exclusive resources, no one else can or will, and each such dereliction is a loss not only to that country itself but to the common heritage.

PEOPLE IN RELATION TO RESOURCES

Group claims
As far as group claims to environmental resources are concerned, one cannot help having personal sympathy with developing countries in their challenge to historical inequities and their insistence that the present gap between the 'haves' and the 'have-nots' is intolerably wide in the light of contemporary resource scarcities. The rich should indeed give a much higher portion of their

wealth, in the form of both economic and technical assistance, to the poorer multitudes of this world. Yet, as the public order theory here would predict, and as history has certainly borne out, it is futile to rely on the inherent generosity of nation-states impelling them to act for the overall good on the basis of the solidarity of humankind.

While urgently promoting multilateral assistance efforts, therefore, developing countries should also seek other practical objectives. Bilateral aid is more likely to be forthcoming, as the benefits to the donor state and the significance of that state's contribution are most readily perceptible; but, from the point of view of the sovereignty and national pride of the recipient, such programs may be politically unpalatable for the same reasons. A vital area commanding the attention of disadvantaged states is, therefore, new resources within the inclusive domain, and, in particular, the rapid establishment of a workable international regime for the exploitation of the deep seabed. Failure to conclude a treaty to this end, with the resultant exclusive appropriation of these resources by those countries possessing the technological capacity to do so, will be a great loss to developing countries (in view of the logical action and past proclivities of advanced industrial states); they may well be at a similar disadvantage with the resources of Antarctica, outer space, and so on. Another good strategy from the point of view of developing countries is clearly to focus on the visibility factor and concentrate their demands on those states whose action is currently under closest international scrutiny, which suggests, among other things, exerting intense political and social pressure on the newly opulent Arab oil-producing nations.

Individual claims
As to resource claims of individuals, regarding the nationality and movement of peoples, the UN Population Conference recommended that governments and international organizations generally facilitate voluntary international movement.[27] It also accepted as a fundamental principle the assertion that the formulation and implementation of population policies is the sovereign right of each nation.[28] To a large extent there is conflict between these two guidelines, as greatly increased voluntarism may undermine the necessity for some governments to decide on policies and devise programs for control of overpopulation, rational resource use, and social progress – in addition, of course, to the fact that a policy of free choice of nationality is demonstrably unacceptable to those nations potentially on the receiving end of the vast influxes of migrants.

Thus, on balance, the goal will perhaps have to be modified to increased voluntarism in accordance with internationally defined standards of human rights. A code would have to be developed taking account of such varied problems as ethnic and religious minorities, refugees and displaced persons arising from wars, forced migrations, migrant workers, 'brain drain,' and other interrelated factors. Until the world community is capable of devising and upholding such an order, countries affected by the outflow or the reception of a significant number of migrant workers should seek to conclude bilateral or regional agreements which would regulate migration, protect the migrants themselves, and further the interests of the countries concerned. *Ad hoc* international agreements or arrangements will also be needed for the relocation of refugees and in various other areas concerned with the 'quality of life'. And in cases of indignation over gross violations of human rights, governments perceiving and desiring to rectify such conditions may bring extraneous economic and social pressure to bear in order to achieve this (as, for example, the United States attempted through making certain trade preferences conditional on the relaxation of Soviet restrictions on the emigration of Jews to Israel a few years ago,[29] and as the Carter administration is now trying to do much more broadly in its human rights program).

Most if not all population policies are precariously balanced compromises, which are immensely imperfect in that they facilitate only the minimum basic requirements of human rights and human dignity and do not seek to promote optimum freedom of individual choice. They have the virtue, nevertheless, of being realistic in recognizing the limitations both of historical patterns of development and of inequalities in contemporary resource distribution, while at the same time giving some concrete effect to the spirit of the Universal Declaration of Human Rights and other expressions of human solidarity and dignity.

In the final analysis, however, neither the goal of increased voluntarism nor other goals of environmental protection, preservation, and enhancement can be made realistic in the absence of effective checks on numbers of people. One thing that is certain from the ecological perspective is that – contrary to the assertion of the World Population Conference that the formulation and implementation of population policies is primarily and almost totally a matter of national concern[30] – the extent of the global population and of certain sectors of it is a matter of utmost common concern. From the global perspective, as was amply demonstrated by the World Food Conference,[31] even with sharing there is not enough food to free all from hunger, and some nations suffer famine while others are still overfed. Yet, if the developing countries have a right to demand more food from the affluent, they have a concomitant responsibility to check their own ever-growing multitudes. And in an even more general sense, although the dynamics of the relationship between population pressures and the proclivities and capacities for aggression are not fully understood, it is clear that overpopulation exerts strong pressures towards an aggressive foreign policy and thereby contributes to the destabilization of international relations.[32] As regards specific population sectors, the grossly overcrowded conditions of developing countries violate human dignity and are therefore a rather abstract concern of international human rights. But changes in the number of people in advanced industrial states have more direct and concrete significance from the environmental point of view, because of the vast per capita consumption of energy and other resources by these countries.

To add a brief word about the specific relevance of law to world population,[33] from earliest times, law – including international law – has been pronatalist. By and large it continues to be so. Several countries have, however, begun to revise their law to allow abortion and sterilization, to limit family allowances, and in some cases even to mandate the official provision of family planning services.[34] This is a start, but it is in addition essential to remove less direct legal impediments to voluntarism – that is, all the anachronistic legislative measures and social practices which discriminate on grounds of sex and relegate women to domestic roles out of the mainstream of socio-economic development.[35] Furthermore, governments have to review and revise the whole complex of their public health, education, and welfare regulations so as to build in incentives for smaller rather than larger families. Related areas, such as property restrictions, domestic relations law, and so forth, warrant contemporaneous re-examination, and religious law is also of obvious importance. Beyond all such encouragement of voluntarism, nevertheless, the case for planned, compulsory regulation of reproduction[36] – as compatible as possible with basic freedom of choice – should at least be given a fuller hearing, if only to reveal its horrendous portents in advance of the absolute necessity to choose between imposing such measures and leaving the matter to the apocalyptic horsemen.

Evolution of international environmental organization

International organization, characterized here in terms of multiple and interpenetrating processes of authority and control ranging from subnational to national to international or global levels, has evolved primarily in a groping, tentative manner with emphasis on specific problem-solving and palpably less concern with doctrinal niceties. This is particularly true of international environmental organization, which is still a largely patchwork system geared towards meeting little-understood needs and constituted in response to highly undifferentiated demands. It is, however, goal-oriented in that it is directed towards the concrete attainment of the expectations and aims of the world community, in this case of international environmental law. Thus, to a great extent, deficiencies in the organizational processes reflect basic weaknesses in the underlying legal order, while these legal weaknesses themselves tend to reflect the relatively primitive stage of development of international society or a sense of world community. In the final analysis, every system of law relies for its enforcement on the will of the political community it purports to regulate, and this is particularly true of international law. But the community itself must have some way of compelling individual adherence to its collective will, and I hope here to offer some basic recommendations as to how this may better be developed as regards environmental norms and standards.

More complete information of all kinds, both basic scientific data on ecological unities and sociological or other social scientific knowledge of political ecological interrelationships, is crucial to any efforts at improving the situation. Such information is fundamental to the evaluation of environmental costs and benefits by any authority and therefore to the rational management of resource allocation and use.

Beyond that, the public order theory here suggests that it is essential carefully to define common international environmental objectives and then to develop some systematic means of compelling states and others to act in accordance with them. In the absence of central coercive institutions, some such enforcement authority should be lodged, in so far as possible, with those states that have an independent national self-interest compatible with the furtherance of the broadly inclusive common good. This idea has already been explored at some length in the discussion of the concepts of custodianship and delegation of powers above. It lends obvious support, for example, to port and coastal state enforcement jurisdiction in treaties concerned with the protection of the marine environment. Some of the more far-reaching implications of the approach in each of the basic functional areas outlined in chapter 4 will be analysed in this section. The incipient trend described earlier towards greater multilateralization, at least in the initial prescription of international environmental law, accompanied by increased reliance on unilateral motivations for its application, if sustained, would represent movement in the right general direction. A lot of specific improvements, however, are desperately needed.

ENVIRONMENTAL INFORMATION
If fully implemented, 'Earthwatch' – with the Global Environmental Monitoring System (GEMS) for gathering and processing environmental information, the International Referral System (IRS) for facilitating its dissemination and exchange, and the International Register of Potentially Toxic Chemicals (IRPTC) for reducing the hazards of chemicals in the environment[37] – can provide the international community with a comprehensive environmental intelli-

gence network. Despite its promise, however, the system could well bear improvement.

First, in addition to biochemical information, the assessment program should seek to incorporate information regarding the social factors that are also so fundamental to the ecological perspective. Social scientists have only just begun to develop 'social indicators' that would enable standard recording of these variables,[38] but some sort of monitoring of social or sociological costs and benefits is critical to any meaningful assessment of policies for the human environment.

Secondly, the program should be expanded to identify and keep track of developments affecting the environment both in international and in national legal systems. The Environmental Law Centre of the International Union for the Conservation of Nature and Natural Resources (IUCN), which is located in Bonn, West Germany, may serve as the nucleus, but the scope and nature of inputs and the geographical and substantive reach of its information system will have to be broadened.

Thirdly, as a basic foundation to all of this intelligence activity, further research should be undertaken to clarify the confusions and difficulties of *The Limits to Growth* and other such studies[39] and to evaluate the general recommendation of 'no-growth' economic strategies in the light of both natural and social conditioning factors. At the present time there is inadequate national, little regional, and virtually no global political-economic planning for the long-term future.

Fourthly and finally, freedom of scientific and social scientific information is important to the successful implementation of Earthwatch and other information programs. Provision should be made for keeping all currently inclusive resources open to scientific inquiry by all and also for establishing the responsibility of states for monitoring environmental variables within their own territory and territorial waters and for sharing and exchanging that information. In the context of the present law of the sea negotiations, states may have both jeopardized and furthered these aims. They have agreed that, in its exclusive economic zone and on its continental shelf, a coastal state will have the right to regulate, authorize, and conduct marine scientific research, and that any research therein shall require its consent; such consent shall, however, be granted 'in normal circumstances.'[40] It can only be hoped that coastal states will live up to the responsibility inherent in this formulation and in normal circumstances further the free and open development and exchange of information. On the more positive side, in a draft article on monitoring, they have also assumed an express responsibility to participate in international monitoring programs.[41] Although there are now quite a few successful ongoing international monitoring efforts on a voluntary basis, this may be the first broadly multilateral treaty in which states assume such a positive and binding commitment to participate.

ENVIRONMENTALIST PROMOTION

One has only to think of the widespread demand for a moratorium on whaling, including its promulgation in Recommendation 33 of the Stockholm Action Plan and in discussions at subsequent sessions of the UNEP Governing Council to recognize how little public opinion alone did for the whales; the pertinent draft article produced by UNCLOS III, for example, at present merely contains the disclaimer that '[n]othing in the present Convention restricts the right of a coastal State or international organization, as appropriate, to prohibit, regulate and limit the exploitation of marine mammals,' without even an explicit reference to an appropriate conservation standard.[42] It is essential to supplement any hortatory solicitations by mobilizing support within relevant arenas, which requires activating effective élites who influence national and consequently international policies and the progressive development of international environmental law. Various citizens groups and other non-governmental bodies are attempting advocacy towards this end, and they should be given greater psychological and financial support in their role as promoters of the public interest (perhaps even by governments, so as to provide some counterbalance to military-industrial pressures). Lawyers, natural and social scientists, scholars, and others who have a special claim to be heard on the basis of their particular expertise ought also to be encouraged to make use of this potential by adopting the role of concerned environmental activists.[43] And when national governments appoint themselves promoters of international environmental causes, they naturally have a unique capacity for directly affecting outcomes.

Overall, international public opinion, although insufficient in and of itself, is not inconsequential. More rather than less money and energy should be channelled into broad-based education, training, and public information. This is not only a form of environmentalist promotion; it is also fundamental to the achievement of the substantive end of improving the quality of life in an environment conducive to human dignity and well-being. Due to the scope and expense inherent in such campaigns in our age of mass participation and mass thrust of communications technology, specially organized public interest groups backed by charitable foundations can and do make an exceptional contribution in this area. International secretariats, because of their relatively internationalist or at least non-nationalist character, also have a special supportive role to play through direct education, training, and public information efforts of their own as well as technical and financial support to national and local programs; it follows that the creation by UNEP of a special Revolving Fund for information activities is a particularly commendable step forward.[44]

INTERNATIONAL ENVIRONMENTAL LAW-MAKING
The solutions to a great many environmental problems are dependent on the formulation of clear and adequate legal norms. The international law-making process, whether customary or conventional, is complex, laborious, tenuous, precarious, and usually painfully slow. Some problems, such as the creation of a regulatory regime for the oceans and other inclusive resources, require active international agreement for their effectuation. In such cases, the recent trend towards agreement by 'consensus' rather than majority or plurality voting is a favourable development,[45] being more reflective of the distribution of effective power and other political factors and therefore more likely to result in a realistic and viable regime. Other changes, such as extensions of exclusive competence over formerly inclusive resources, can achieve legality over time either by way of actual agreement or through the passive acquiescence of the general international community. Consequently, at some relatively early stage of policy-making on international law issues, certain actions have to be considered outside the law until or unless they attain acceptability to or condemnation by a sufficient number of states. But such validation or prohibition may often, nevertheless, be acquired more rapidly than could be accomplished by multilateral negotiations towards the conclusion of conventions. Thus, it makes little sense to try to draw a sharp distinction between unilateral and multilateral action *per se* or between national policies and international law. What goes on is a process or a number of interrelated processes of action and reaction, whereby international law is responsive to the will of nations and becomes the embodiment of community expectations. It is much more important to evaluate its substance from the perspective of the fulfilment of common interests than to argue over the procedures and methods by which it is prescribed (except, of course, in so far as the form itself dictates the substance).

The Stockholm Conference clarified a lot of customary expectations, and the UN General Assembly has since reinforced many of these expectations. Environmental policy has also been crystallized in a number of recent treaties and other international agreements. The prescriptive order nevertheless remains incomplete, inconsistent, and inadequate. First of all, some examples of areas where law is non-existent or ambiguous are land-based sources of marine pollution generally, the conservation of high-seas fisheries, marine mammals and weather and climate modification activities. But there are many theoretical and practical gaps and discrepancies which are not so readily apparent, and some body of experts should be commissioned to make studies – as the International Law Commission once did for the law of the sea[46] – of potential environmental norms and to come up with recommended standards of behaviour and action. At its second session, the Governing Council of UNEP recognized this need and authorized the Executive Director to convene informal working groups of legal experts to advise him on how best to contribute to the future development of international environmental law; and a further decision of its third session gave him a clear and broad mandate to take active measures towards the sound development of environmental law world-wide, upon which work has already been initiated by several UNEP working groups.[47] But even such unexceptional decisions could not be achieved without overcoming some formidable opposition from certain states, so the international community still has a very long way to go in attempting to establish effective patterns of environmental law-making.

Furthermore, where conventions have already been concluded, states should be urged, perhaps by reciprocal pressure, to accelerate their national procedures for bringing them into force. The 1969 IMCO 'Public' and 'Private Law' Conventions, for example, took until mid-1975 to achieve sufficient numbers of ratifications, and the 1969 amendments to the 1954 Convention for the Prevention of Pollution of the Seas by Oil did not finally enter into force until January 1978.[48] At the present time, international conventions relating to maritime pollution not yet in force include the 1971 amendments to the 1954 Oil Pollution Convention, the 1971 International Convention on the Establishment of an International Fund for Compensation for Oil Pollution Damage, and the 1973 International Convention for the Prevention of Pollution from Ships.[49] The IMCO Fund Convention will, however, enter into force before the end of this year. Moreover, international agreements in general may ultimately, through widespread international acknowledgement and customary behaviour, become common community expectations or generally accepted international norms and standards – that is, 'law' – even as regards non-signatories or non-ratifiers. Still, having managed to reach agreement and define commonalities of interest, states should rapidly reaffirm their commitments domestically through acceptance and ratification and by conforming their national laws to these and other internationally agreed obligations.

On a related matter, a 'tacit amending' procedure has been proposed in IMCO and incorporated in several new conventions. The basic idea is that failure on the part of a state to signify dissent within a specified period of time is to be taken as acquiescence, and accordingly amendments to certain aspects of IMCO conventions are enabled to come into force in an accelerated manner.

Such a procedure has already been incorporated in the Pollution from Ships Convention, the 1973 Protocol to the 1969 Intervention Convention, and others.[50] It should certainly help to accelerate part of the law-making process.

INVOCATION FOR ENVIRONMENTALIST CAUSES
The main recommendation as regards the invocation function is recognition of the standing of any state to advocate inclusive community interests before international tribunals and in other international arenas.[51] In fact, of course, a state is likely to exert itself and invoke international machinery only when its own national interest happens to coincide with the common good or for propagandistic purposes, but that should in no way be taken to derogate from the importance of states being authorized to perform this role. In *Nuclear Tests*, for example, if the World Court had specifically considered the claim of the applicants to be representing inclusive common interests in addition to their own exclusive interests, might it not have decided to issue a declaratory judgement instead of merely finding that the sole objective of Australia and New Zealand – that is, termination of the specific series of tests – had been fulfilled?[52] A further idea deserving of consideration is the provision for an international ombudsman charged with invoking processes or intervening in them as a representative of inclusive concerns when the common environment is threatened.[53] Whether this be the Executive Director of UNEP or some other individual or organization, the ombudsman would have to be directly connected with the Earthwatch intelligence facilities and supplied with a competent legal staff.

With regard to both inclusive and exclusive resources, it is also generally recommended that non-official actors be accorded greater access to relevant arenas. States are the primary but certainly not the only actors on the international scene, and many types of environmental situation are more economically and expeditiously treated – at least initially – as transnational rather than international problems. On the international plane, short of a change in the Statute of the World Court, this suggestion can be furthered both by an increased willingness on the part of national governments to stand forth in a representative capacity for public interest groups and other environmental claimants and by the provision of alternative arenas open directly to private citizens and non-governmental actors of various types. The latter should be particularly useful in cases of transboundary pollution and other alleged injury to private interests, as redress through international claims commissions tends to be such an interminable process. On the national level, through development of uniform laws reciprocally between states and by particular countries individually, what is needed is the relaxation of standing requirements before courts and administrative tribunals to accord greater recognition to groups or individuals presenting transnational environmental claims. The 1974 Nordic convention, which grants to citizens of the four countries concerned reciprocal access to each other's courts on the same basis as nationals of the host country in environmental suits,[54] is an admirable model in this regard.

APPLICATION OR ENFORCEMENT OF
INTERNATIONAL ENVIRONMENTAL LAW
As far as application or enforcement of international environmental law is concerned, the authority of particular states to make applications of international standards (and, in their absence, of reasonable national standards) appropriately clarified should be recognized and extended. This is especially so, again, where the applying state has an exclusive interest highly compatible with and in furtherance of widely inclusive common interests – as in the cases of the United States' port state program and Canada's Arctic waters pollution act for the oceans,[55] and Australia and New Zealand's attempts to prevent radioactive contamination of the atmosphere in South Pacific areas.[56]

More generally, the public order theory developed here leads one to support such innovations as both port and coastal state enforcement jurisdiction for international law in regard to protection of the marine environment. The right of the flag states to apply the law to their vessels should also be maintained in order to uphold traditional doctrines of freedom of the seas, but they cannot logically be expected voluntarily to start exerting great efforts on behalf of the common good probably at the expense of their own immediate, short-term national interests. Port states, however, have an independent interest in the safety and order of their ports and offshore installations and therefore an incentive to apply strict standards; yet this is somewhat balanced by their competing interest in promoting and enlarging their international trade and use of their ports. Coastal states, in so far as they are also ports of entry, have the same duality of interest; but in so far as there is a question of traffic that may endanger their environment and the marine environment transitting their coasts, their impetus for voluntarily enforcing pollution prevention and other environmental protection measures is most nearly comparable to inclusive community interests. The rule of reason will, of course, have to apply, so as not to open up the system to perversions and outright graft; and compulsory dispute settlement provisions might also be of some ancillary use to this end. Still, overall, the best recommendation from the ecological perspective remains to delegate the power to enforce the law to those states in whose own self-interest it is to do so as custodians of vital community interests.

When disputes arise concerning the interpretation or application of international environmental law, it is essential that procedures be provided for their settlement in order to safeguard against abuse of rights and powers by any state and to ensure that measures taken by the custodial state correspond to the interests of the international community and are carried out in accordance with its guidelines. At times differences of opinion can readily be solved through interstate negotiations, but in other cases non-state appliers will have to be introduced, and their role ought also to be acknowledged and enhanced. A Special World Environment Court seems far from being called for at this time and may indeed never be found advantageous. The International Court of Justice can, however, be streamlined in ways already envisioned in its Statute (through the use of chambers, assessors, and so on),[57] and other traditional third-party dispute settlement procedures (good offices, mediation, conciliation, arbitration, judicial settlement in other international and national courts and administrative tribunals) can be tailored more routinely and accurately to take account of and to give greater weight to environmental factors in reaching their decisions.

TRANSITIONAL MEASURES
Problems of transition are bound to be encountered in the evolution of a more ecologically sound world public order. The biggest problems with regard to reducing the costs of change occur with respect to developing countries. At Stockholm and elsewhere it has been repeatedly emphasized that in the developing countries most of the environmental problems are caused by or related to underdevelopment. The efforts of advanced industrial states to combat the ills caused by modern technological development should not, therefore, be allowed to preclude the economic and social progress of the poorer nations.[58] The short-term special provisions to accommodate the legitimate expectations of developing countries may result in the creation of 'pollution havens' for industry and other temporary undesirable side-effects. Such relaxation should not, however, extend to the lowering of environmental protection standards for inclusive resources below international minimums, thereby destroying the potential for future development.[59]

Sometimes the legitimate expectations of private interests within particular countries will also warrant compensation as a result of changes in the international public order. Such matters, however, are clearly subsidiary to broader policy considerations in the creation of a viable world public order, and they must almost necessarily be left to national determination after the basic international framework has been established. In this connection, the US 'Metcalf Bill'[60] and its successor deep-sea mining bills have been justly criticized for their somewhat insidious attitude and as an unjustifiable attempt to stake out a unilateral claim to nodule-mining in international waters, tending to subvert contemporaneous efforts by the international community to work out an acceptable inclusive regime for these non-renewable resources. It may, however, be in accordance with traditional US and general state practice on spreading losses and thereby reducing some costs of change to provide for investment guarantees or compensation to US miners for losses incurred in pursuance of activities encouraged by the US government, and thus the legislation may have been unfairly attacked in this regard. Whether such guarantees are necessary and proper, nevertheless, has yet to be determined by domestic political processes and national constitutional law in the light of all relevant factors, about which no opinions are offered here.

APPRAISAL OF ENVIRONMENTAL WORLD PUBLIC ORDER
It is too early to expect major transformations as a result of the contemporary upsurge of environmental problems in the world community. Certain very important changes affecting the capacity of the international system to deal with environmental costs and benefits are, however, already apparent. The advent of exclusive economic zones on a functional basis, the move towards the establishment of state responsibility for environmental injury, and the numerous treaties attempting to set up a regulatory regime for the protection and preservation of the marine environment are some important examples. But there must be follow-through evaluation of the extent to which these developments actually do evolve towards a more ecologically sound legal order and an ecological approach to resource management.[61]

The United Nations Environment Programme has already begun appraisal of the functioning of the UN system from the ecological perspective, and this overview is being expanded to cover the public order as a whole. In the contemporary international system, states collectively and individually will have to bear primary responsibility for self-appraisal and correction of their actions in protecting inclusive and exclusive resources, but they may be spurred on and assisted by the information-gathering and assessments of a committed but relatively detached observer such as UNEP. The behaviour of national governments must, of course, continue to be the subject of reciprocal observation by other states in international environmental bodies and elsewhere; and public interest groups, universities, foundations, private scholars, and concerned citizens also have to shoulder responsibility for an ongoing appraisal function.

Watershed reappraisals, too, have their place. While they often seem an excessively cumbersome and expensive way of reaching decisions, general international conferences do refocus attention and serve as a benchmark for evaluating progress or the lack thereof. On balance, therefore, in addition to advocating continuing review, support should be given to the convening of a Second United Nations Conference on the Human Environment in the not too distant future.[62]

FUNDING

The most important point to be made on the matter of funding is that much more is needed, which plea has been heard before and should continue to be sounded. The United Nations itself lacks anything more than a minimally effective means for making compulsory budgetary assessments, and the major proportion of funds for UNEP and other international environmental programs and activities must come from voluntary contributions. The great reluctance of states to contribute voluntarily has been pointed out many times, and there is little that can be recommended in this regard other than to keep pointing out the potential benefits to be gained and to advise the exertion of whatever social and political pressure is available – which may sometimes be substantial. Also, in many instances, international environmental protection measures can be achieved by using international funds as a catalyst for national or other efforts. And it goes without saying that it would be an enormous help in raising funds if UNEP and other environmental organizations effectively demonstrated that they can both use such funds wisely and advantageously and conduct their operations in accordance with sound fiscal management practices and procedures.

One minor suggestion for supplementary international financing is that some of the revenue (if any) from the exploitation of the international sea-bed area should be allocated directly to the maintenance of the Sea-bed Authority and perhaps of other international institutions and their programs. Suggestions along these lines have been offered in other contexts, such as the renewed 'Horowitz Plan' put forward at Stockholm that slightly more than 3 per cent of any special drawing rights created by the International Monetary Fund be used to promote action for environmental amelioration.[63] What any such proposals have in common is that they seek to allocate 'new' resources rather than asking any country to give specifically of its own. Sea-bed revenues would presumably have the advantage of being in a relatively well-known and readily understood form (that is, leasing payments), by comparison to the slightly esoteric nature of special drawing rights, but the basic theory is the same. Should UNCLOS III in the end fail to achieve agreement on a regime for the international Area, it should be realized that not only developing countries but the community as a whole will lose if sea-bed resources are by default thrown open to exclusive appropriation by those states with the most immediate technological capacity for their mining.

Review of the public order theory and concepts

Throughout the present work, I have viewed the world public order as a rather loosely organized system of multiple and interpenetrating processes of authoritative decision. States are the primary actors and behave substantially analogously to private individuals in a market economy in seeking to maximize their value positions. States possess common interests, many of them critical; but they cannot be relied upon consistently to act voluntarily in furtherance of the common good, unless their own national interests coincide or can be made to coincide with broadly inclusive interests. Yet under certain circumstances they may be compelled to do so for the benefit of the total community. Three main strategies have been identified as likely to be effective towards these ends, as means of giving effect to previously unaccounted for environmental factors: first, internalizing the environmental cost-benefit calculations in regard to various resource uses to a single state with authority and control over them; secondly, determining as specifically as possible the environmental costs of certain other acts or activities and holding states responsible for them in their interactions with each other; and thirdly, direct collective regulation with effective enforcement provisions in respect of still other activities in order to take better account of environmental costs and benefits. To some extent there has been movement along all three of these lines of development, the choice of means being greatly influenced by the nature of the resources being regulated. But the task remaining is enormous.

This is admittedly a somewhat imperfect characterization of the contemporary public order. Distortions are introduced both by the bureaucratic politics of nation-states themselves and by the fact that certain activities of some other actors (from explorers to multinational corporations to acts of piracy) have occasionally managed to elude the effective control of any and all sovereign states. The abundance of data on observed trends in international environmental law and organization has, nevertheless, strongly supported the hypotheses about states' behaviour that can be drawn from my theoretical perspective. I have therefore not found it necessary to expound upon other useful but more extreme bodies of literature – on the one hand, the economic theory of the creation of property under conditions of anarchy,[64] and on the other, discussions of optimal resource allocation under a highly organized and centralized system of government.[65] Both of these and several other clusters of theories have pronounced utility under certain existing circumstances, but limitations of time and space have impelled me to focus on the overall patterns, overlooking certain relatively minor exceptions and variations characteristic of contemporary international behaviour. My main concern, to repeat an initial observation, has been to provide a comprehensive description of the development of international environmental law and organization to date and to present and analyse basic policy alternatives for the future available within the context of the present international system.

I have made recommendations couched in terms of concepts of custodianship and delegation of powers as set forth by Canadian diplomats in law of the sea negotiations. In so far as these terms mean what I have defined them to mean, they provide a viable conceptual framework for the future development of environmental world public order. They have not been highly specified or elaborated upon by their sponsors, and I have no intention of being automatically committed to everything Canadian or any other policy-makers may wish to read into them (or, for that matter, may already have incorporated). The terminological borrowing here is primarily to demonstrate that proposals in accordance with my basic theoretical perspective and policy suggestions have already been advanced and are being advocated in international diplomacy and also to give credit where credit is due. Several other countries besides Canada have formulated and supported policies along the same lines, albeit without the neat doctrinal encapsulation, and they are equally deserving of consideration and support.[66] What I have tried to do, in sum, is to take a skeletal form and give it some flesh and substance, in order to clarify the criteria by which decision-makers should be guided.

Perhaps Canada itself will take occasion to clarify this concept and to codify and progressively develop international environmental law in the near future. As this book goes to press, the world has recently been dramatically reminded of environmental dangers by the fiery crash of Cosmos 954 – a Soviet spy satellite carrying a nuclear reactor to power its ocean-scanning radar and radio circuitry. The nuclear-laden satellite crashed into the atmosphere over a remote Canadian wilderness area, but had it failed one pass later it would probably have plunged towards earth near New York City at the height of the morning rush hour. United States and Canadian experts jointly undertook a massive search, christened 'Operation Morning Light,' to detect hazardous radiation and satellite debris. The total costs of the search, which did find high levels of radiation and some small fragments from Cosmos, exceeded $5 million and may have reached $10 million.[67] This incident raises questions of interest differentiation, treaty interpretation, customary norms, and many of the other topics discussed throughout this book – for example: How should the public order resolve the clash between common and special interests involved? How are the Outer Space, the Liability and Compensation for Damage Caused by Space Objects, and the Rescue and Return of Space Objects treaties to be interpreted and applied as regards coverage for the search? What responsibility was there to warn of this incident, and was it fulfilled? And are the risks involved acceptable, or should there be an absolute prohibition of nuclear material in space absent 'fail safe' systems? The questions could go on at length. The central point is, however, that here is practically a textbook case: ultrahazardous activity results in the crash of nuclear materials in one of the most environmentally conscious countries on earth, which is aided in dealing with the situation by another such state. From the perspective of the development of an ecologically sound world public order, what are the various parties involved going to do about it?

A final word to political philosophers

As was emphasized in the introductory section, the goal here is not optimality in the traditional economic or political sense. This work was written from the perspective of a scholarly observer and analyst looking at *this world* at *this time* and seeking means for achieving a minimum environmentally sound future world public order. I have been concerned throughout with assessing progress and evaluating the efficacy of environmental strategies from the point of view of their recognition of ecological unities and the furtherance of environmental goals in the light of the limitations inherent in the existing political ecological context. Some alternative approaches may further transition to different global political constellations more than others, but I have stopped short of choosing among them on these grounds. Whether or not centralized world government of any type is preferable to the current pluralistic system cannot be decided on ecological grounds alone, once one rejects the hypothesis of inevitable ecocatastrophe under the current order in the near future. These vital questions of the nature of an optimal or utopian world public order and how to get there are matters for political philosophers extending way beyond the aspirations of the present inquiry. I wish them well, but for the meantime end by repeating the famous conclusion of that early environmentalist Candide: 'Il faut cultiver notre jardin.'

PART FOUR

**Problem Studies in
International Environmental Law and Organization**

6

State responsibility for environmental protection and preservation

Given emerging ecological and political ecological interdependencies and in view of the nature of the contemporary international system, it follows that there is a basic obligation upon states to protect and preserve the human environment, and in particular to use the best practicable means available to them to prevent pollution and other destructive impacts on both exclusive and inclusive resources. This basic obligation may seem obvious and self-evident, but the notion represents a radical departure from traditional *laissez-faire* doctrines of state sovereignty in general and in regard to specific inclusive resources such as the seas. Although it is fundamental to the United Nations Declaration on the Human Environment and to many emanations from customary international law, at present there exists no explicit treaty obligation laying down this responsibility in comprehensive terms capable of effective implementation. In fact, the whole international law of state responsibility for environmental protection and preservation and complementary questions of liability and compensation for environmental injury is currently in a state of obscurity and flux, and it is hoped that the present discussion can make some contribution to its codification and progressive development.

In their interactions with each other in the use and enjoyment of resources, states and other actors on the world scene make claims and counter-claims for the prescription and application of authority. The first consideration to be faced by international processes of authoritative decision in evaluating these contentions is the permissibility of the activities in question at all – that is, whether or not they have been prohibited by the world community. Assuming that the activities are permissible in some degree, the problem then becomes managing the public order consequences of the resulting deprivations so as properly to take account of environmental costs and benefits. The latter objective may be sought through measures to deter actors from incurring environmental risk and/or through allocating compensatory costs for the *ex post facto* control or elimination of injurious results. In keeping with the basic aim of international tort law to minimize unauthorized coercions and deprivations of all kinds, the fundamental question here is how to shape a law that serves all these purposes.

The aims or objectives being sought through what is very loosely termed the law of 'state responsibility' for environmental protection and preservation, it follows, can be usefully described in terms of three broad subgoals: prevention of environmental deprivations, deterrence of impending environmental harm, and reparation or compensation for environmental injury which nevertheless results. After some initial observations on the general background of the doctrine of state responsibility in the environmental context, I will examine international environmental law in the light of each of these policy subgoals served by sanctioning processes ('sanctioning' being used in its broad sense of that which gives force to the law, including penalties or other punishment for breaking it). The intention here is not only to explore the relevant substantive legal principles, but also to try to shed some light on the practical nature of the problems involved and of the procedures necessary to cope with them.

Background of state environmental responsibility

Less than a century ago it was possible for US Attorney General Judson Harmon without embarrassment to assert that '[t]he fundamental principle of international law is the absolute sovereignty of every nation, as against all others, within its own territory' and therefore to conclude that 'the rules, principles, and precedents of international law impose no liability or obligation' which inhibits a state from using the resources within its territory as it chooses without regard to the impact upon others.[1] Fortunately, this ultranationalistic notion was not adhered to at that time, nor has it won sympathy in the subsequent development of international law. Instead of the Harmon doctrine, the international community has clearly adopted the maxim of *sic utere tuo ut alienum non laedas* (use your own property so as not to injure that of another) fundamental to both Roman and common law. As a result, today state responsibility can be regarded as 'a general principle of international law' or as 'a concomitant of substantive rules and of the supposition that acts and omissions may be categorized as illegal by reference to the rules establishing rights and duties.'[2]

'State responsibility' has traditionally been defined in narrow terms of a wrongful act or omission which causes injury to an alien, but the doctrine has not been confined to that context. As even the reporters of the Harvard draft Convention on the International Responsibility of States for Injuries to Aliens have acknowledged, '[t]he responsibility of a State may also be engaged by a violation of any treaty or any rule of customary international law under such circumstances that no injury to an individual is involved.'[3] Since such responsibility may be either original with the state itself or vicarious as a result of unauthorized acts of its agents or nationals, there are a great number of possible permutations and combinations of international claim situations.[4] Problems may also arise in connection with the participation of several states in the same act, and they can be further complicated by questions of the responsibility of one state for the acts of another state or its nationals.[5] This whole polemic has for several years been under discussion in the UN International Law Commission; and it is interesting to note that in a quite recent draft on state responsibility the ILC said that an 'international crime' may result, *inter alia*, from 'a serious breach of an international obligation of essential importance for the safeguarding and preservation of the human environment, such as those prohibiting massive pollution of the atmosphere or of the sea.'[6]

Meanwhile, *state responsibility* has acquired a more expansive meaning, and the term is now commonly employed by international lawyers to encompass a broad range of conditions under which international obligations may be incurred. In the environmental context, there has been explicit acceptance of the principle that states must bear responsibility for the effects of their actions on the environment of other states or the common environment. The international community has also expressed concern for clarifying the circumstances under which violation of a substantive norm entails an obligation to make reparation or to pay compensation for any resultant damage. Principles 21 and 22 of the Stockholm Declaration, which have frequently been cited in the present study and elsewhere, embody the current community expectations, and it seems worth repeating them fully in juxtaposition here:

Principle 21
States have, in accordance with the Charter of the United Nations and the principles of international law, the sovereign right to exploit their own resources pursuant to their own environmental policies, and the responsibility to ensure that activities within their jurisdiction or control do not cause damage to the environment of other States or of areas beyond the limits of national jurisdiction.
Principle 22
States shall co-operate to develop further the international law regarding liability and compensation for the victims of pollution and other environmental damage caused by

activities within the jurisdiction or control of such States to areas beyond their jurisdiction.[7]

Logically understood, this formulation would have to incorporate responsibility for preventing irremediable or noncompensable effects as well as liability for actual damage, responsibility for warning of reasonably foreseeable environmental consequences of otherwise lawful activities, and responsibility for submitting to peaceful and expeditious settlement disputes related to any of these matters. There has been some subsequent elaboration along these lines, but many crucial issues remain unstated and substantially misunderstood.

With reference to this expansive notion of international obligations, the pages that follow should indicate what progress has in fact been made towards developing effective sanctioning processes for the implementation or enforcement of international environmental law. I shall also indicate areas where there are basic deficiencies in the evolution of a regime of state environmental responsibility.

Prevention of environmental deprivations

Within the overall realm of state environmental responsibility, the first objective of international sanctioning or implementing processes is that of the long-term prevention of environmental losses. Prevention embraces a great variety of measures and activities designed, over a varying range of time, significantly to reduce the probability of undesirable environmental effects. The instruments of policy involved encompass the whole range of diplomatic, economic, ideological, and military strategies available for the maintenance of international public order. Specific examples of likely 'police action' may include the promulgation of standards and criteria for use and enjoyment, inspection and monitoring of compliance, the prohibition of enjoyment of the resources to non-compliant users, and legal or administrative proceedings to investigate complaints and determine appropriate penalties for violators.

For a comprehensive view, it is necessary to look both at national policing systems and their sanctions for infringement of national environmental protection legislation and at the conditions under which and the means by which these and certain other sanctioning processes are directed towards upholding multilateral prevention prescriptions. Then, with reference to how to determine what new preventive measures may be necessary or advisable in the future, a word ought to be added about the crucial role of monitoring in enabling management of resources according to sound scientific principles.

NATIONAL ENVIRONMENTAL POLICING SYSTEMS

To take as an example what should by now be a familiar piece of legislation, the Arctic Waters Pollution Prevention Act is a model attempt by one state to provide comprehensive environmental policing for its designated area of coverage.[8] No one doubts the particularly severe climatic conditions or other special ecological circumstances of Arctic regions, and the political ecology of the Arctic is exceedingly complex, difficult, and inconducive to joint action.[9] Consequently, taking cognizance of its 'responsibility for the welfare of the Eskimo and other inhabitants of the Canadian arctic and the preservation of the peculiar ecological balance that now exists in the water, ice and land areas of the Canadian arctic,'[10] (and of peculiar political imbalances as well), the government of Canada has gone ahead on its own with this legislation. The act prohibits and prescribes penalties for the deposit of 'waste' in Arctic waters or on the islands or mainland under conditions that such waste may enter Arctic waters. It provides for civil liability resulting from such deposit on the part of persons engaged in exploring for, developing, or exploiting the natural resources on the land adjacent to the Arctic waters or in the submarine areas below the waters, or by persons carrying on any undertaking on the mainland or the islands of the Canadian Arctic or on Arctic waters, or by owners of ships navigating within Arctic waters and owners of the cargo of any such ship (which liability is absolute and not dependent upon any proof of fault or negligence). In addition to these basic substantive provisions, the act empowers the Governor General in Council to make regulations relating to navigation in shipping safety control zones and to prohibit any ship from entering such zones unless it meets the regulations concerning hull and fuel tank construction, navigation aids, safety equipment, pilotage, icebreaker escort, and so on; he may also order the destruction or removal of ships in distress which are depositing waste in Arctic waters or are likely to do so. The same legislation furthermore makes arrangement for Pollution Prevention Officers who among other things may, with the consent of the Governor General in Council, seize a ship and its cargo anywhere in Arctic waters within the hundred-mile zone of coverage or elsewhere in the territorial sea or the internal or inland waters of Canada when there is suspicion, on reasonable grounds, that the provisions of the act have been contravened by the ship or by the ship or cargo owners. Upon conviction for such an offence, a court can order the forfeiture of both the ship and its cargo.

Other nations have made differing policing arrangements in accordance with what they understand as the nature and scope of their pollution responsibility for their own waters and coasts and in regard to inclusive resources. Some of the provisions of the US Ports and Waterways Safety Program have, indeed, already been described elsewhere.[11] It has also been pointed out that flag states in general have, as might be expected, expended very little energy in developing capabilities for pollution prevention on the part of their vessels (which may not even come near their shores).[12] Rather than dwell on the extent or jurisdictional bases for enforcement regimes, however, the point for present purposes is to highlight the basic nature of the policing and sanctioning processes themselves. As was shown by the example of the Arctic waters legislation, a comprehensive set of preventive measures would encompass regulations for the conduct of operations in regard to the use and enjoyment of ecologically interdependent resources on a functional basis, investigatory powers to oversee compliance, subpoena and other powers for the production of witnesses and documents, seizure powers and/or bonding arrangements for ensuring collateral with which to satisfy an adverse judgment, juridical or administrative procedures to evaluate conformity or non-conformity, and civil and/or criminal penalties for inadvertent or wilful violation. The US Ports and Waterways Safety Act even goes as far in its criminal penalties as a provision for imprisonment of up to five years as an alternative or in addition to fines.[13]

The references so far have been drawn primarily from the area of marine environmental protection, since that is the subject of most immediate international environmental concern at the present time. It goes without saying, however, that states individually and collectively also have an interest in protecting their air, rivers, lakes, and other resources from pollution and other kinds of environmental deprivations and their people from various additional nuisances (for example, noise) as well. It would be futile to attempt to catalogue all sorts of national environmental legislation and its enforcement provisions here (although that task has already in fact been begun by others elsewhere).[14] There are some variations, but the skeletal enforcement patterns are not fundamentally different in kind, and for the moment it should suffice to grasp the rudiments of what is needed.

SANCTIONS TO ENFORCE
INTERNATIONAL PREVENTION PRESCRIPTIONS
The above examples have dealt with preventive measures to enforce national legislation, but the requirements are not dissimilar as regards either the implementation or the enforcement of international law. It has been alleged in purist terms that, since in jurisprudence a law is said to have a sanction when there is a government which will intervene if it is disobeyed, 'therefore international law has no legal sanction.'[15] Except perhaps in the most narrow legalistic ratiocinations, that statement is meaningless or irrelevant, since international law has at its service a whole range of political, economic, social,

and even military sanctions on the part of nation-states to secure implementation.

As far as intervention by international machinery itself is concerned, a declaratory judgment by the International Court of Justice or some other international tribunal may have the character of a 'sanction' or 'measure of satisfaction.'[16] The declaration in the *Corfu Channel* case of the illegality of the British minesweeping 'Operation Retail' provides a textbook example.[17] There has, however, been some question about the role of declaratory judgments in cases of potential environmental injury, with at least one writer ruling out resort to them as a means of determining reciprocal rights and duties of parties to a dispute involving extraterritorial environmental interference caused by a *per se* lawful activity in the absence or in advance of actual, provable injury.[18] Yet, as others have pointed out, the arbitral tribunal in the *Trail Smelter* case did, nevertheless, hold that 'it is the duty of the Government of the Dominion of Canada to see to it that this conduct [the future operation of the smelter] should be in conformity with the obligation of the Dominion under international law as herein determined';[19] so the question at the very least remains open.

Unfortunately, clarification has not been provided by the recent *Nuclear Tests* decisions, where the International Court of Justice chose not to confront the problem of illegality.[20] It will be remembered that the ICJ held, in effect, that cessation of the atmospheric testing by France, together with French public statements announcing an intention hereafter to test only underground, rendered moot the issue upon which it had been asked to pass judgment by providing the relief the parties wanted. More specifically, having found that 'the original and ultimate objective' of the applicants in these cases 'was and has remained to obtain a termination of those tests [above ground]' and therefore that the claim 'cannot be regarded as being a claim for a declaratory judgment,'[21] the majority concluded that the controversy 'no longer has any object and that the Court is not called upon to give a decision thereon.'[22] Dissenting judges insisted that not only did this assertion fail to take account of the purpose and utility of a request for a declaratory judgment at international law, but it also changed the scope and nature of the formal submissions.[23] Still, the majority decision was not to decide.

In any event, in the overwhelming majority of circumstances, responsibility for the enforcement of international environmental law so as to prevent injury will rest with individual states. When states ratify international treaties and agreements, they quite routinely have to pass detailed implementing legislation (although some of them are, of course, self-executing). Principle 4 of the

General Principles on Marine Pollution has expressly provided that '[s]tates should ensure that their national legislation provides adequate sanctions against those who infringe existing regulations on marine pollution';[24] and, for another example, each contracting party to the Paris Land-Based Sources Convention has specifically undertaken 'to ensure compliance with the provisions of this Convention and to take in its territory appropriate measures.'[25] But the responsibility of states to conform national legislation and enforcement action to agreed international norms is usually regarded as implied under both conventional and customary international law and is rarely so explicitly stated. The specific sanctioning regimes often vary considerably in accordance with domestic constitutional criteria, although certain policing measures have been the subject of international accord. An example of the latter category might be provided by the Oslo and London Ocean Dumping Conventions,[26] which stipulate that states should set up 'special permit' systems for certain grey-listed substances taking account of various specified factors; but as to actual regulations and enforcement details, these conventions too leave a great deal to the legislative, executive, and judicial imagination of national governments.

On the core issue of what is to be prevented, by whom, and to what extent, a multiplicity of existing treaties and customary obligations were discussed in chapter 3. Just a partial listing of the conventional prevention prescriptions mentioned is long: the 1954 International Convention for the Prevention of Pollution of the Seas by Oil with its subsequent amendments and a whole host of related legislation; the 1971 Oslo Convention for the Prevention of Marine Pollution by Dumping from Ships and Aircraft and the following 1972 London Convention on the Prevention of Marine Pollution by Dumping of Wastes and Other Matter; the International Convention for the Prevention of Pollution from Ships; the Paris Convention for the Prevention of Marine Pollution from Land-Based Sources; the Outer Space Treaty; the 1964 Treaty Banning Nuclear Weapons Tests in the Atmosphere, in Outer Space and Under Water and the later Non-Proliferation, Seabed Denuclearization and Tlatelolco treaties; the Antarctic Treaty with its prohibitions, and other agreements to prevent environmental degradation, up to and including the new Convention on the Prohibition of Military or Any Other Hostile Use of Environmental Modification Techniques.[27]

Most of these treaties provide for action by the national state of those carrying out the polluting activities in question. But, as was also discussed previously, a new and desirable development from the environmental point of view is the provision for coastal state enforcement in the London Dumping Convention and elsewhere and its elaboration at the Law of the Sea Conference.[28] The extent of action provided for is usually the fullest extent necessary to accomplish the specified purposes according to agreed standards and criteria, and the reach of allowable enforcement measures is not usually an issue. In the debate over coastal state enforcement in the deliberations on the law of the sea, however, there has been sharp and highly vocal disagreement over whether or not such jurisdiction should extend to shore-based or in-port investigation and action only, the detention of vessels, stopping and inspecting ships in territorial waters or the economic zone, the power to order them out of coastal zones, the arrest of offending vessels, the institution of proceedings with only monetary or more drastic penalties, and so on.[29] Again from the perspective of environmental protection, full enforcement authority is warranted, qualified only by requirements that such powers be exercised reasonably and non-discriminatorily, without undue interference with other legitimate uses of the seas, and subject to readily available dispute settlement provisions. Since the issues of flag/port/coastal state jurisdiction were discussed in connection with custodianship and delegation of powers in chapter 5, and since UNCLOS III is currently in the process of finalizing a highly detailed set of draft articles on protection and preservation of the marine environment, we can now move on to topics that introduce fresh considerations.

ENVIRONMENTAL MONITORING
Before passing on to the controversial subject of environmental deterrence measures and to the scarcely considered subject of liability and compensation regimes, however, it should be strongly re-emphasized that the acquisition of accurate and broad-based information is critical to the prevention of environmental injury. An adequate international monitoring system is essential if the effectiveness of existing provisions and the necessity for further preventive measures are to be assessed. The need for ongoing and comprehensive monitoring on a global scale was, of course, recognized at the UN Conference on the Human Environment, and the need for monitoring was referred to in at least twenty-three specific recommendations dealing with very diverse areas of concern.[30] Consequently, planning got under way for the establishment of the Global Environmental Monitoring System and an International Referral Service.[31] By mid-1977, the Executive Director of UNEP was able to state as a practical goal the rendering fully operational of both GEMS and IRS by 1982.[32] It is certainly to be hoped that this goal can be met, and that states will continue to update and elaborate their monitoring efforts thereafter.

In the first place, states which permit or engage in activities resulting in the release of substances that may damage the environment have a responsibility to determine the effects of such activities which should be given explicit recognition, in the best interests of their own citizens as well as to be consistent with the rights of other states. Secondly, the system cannot work properly to protect common interests unless there is wide dissemination of information about the release of substances or the known presence of substances likely to cause pollution or other problems either alone or in combination with other effects. Participation in GEMS and other international monitoring systems is at present voluntary only, but at least in one important area – marine environmental monitoring – states have agreed on certain positive commitments. In the UNCLOS III draft article on monitoring states have obligated themselves both to 'observe, measure, evaluate and analyse, by recognized methods, the risks or effects of pollution of the marine environment' and to 'provide at appropriate intervals reports of the results obtained ... to UNEP or any other competent international or regional organizations, which should make them available to all States.'[33] But this is only part of the overall requirement. Similar obligations should be acknowledged and upheld as regards other areas and ecological interdependencies.

Some of these issues are discussed later in connection with environmental warning and notification on pages 159–62 below. The point here is simply to reiterate that the responsibility for acquiring knowledge and assessing the consequences of ongoing and proposed actions must be not merely a time-to-time affair but a continuous individual and reciprocal responsibility of states if the global environment is to be protected. A reliable and up-to-date data base is fundamental to wise and informed political choice.

Deterrence of impending environmental harm

Supplementary to their overall interest in long-term prevention of environmental deprivations, states must also be concerned with the deterrence of particular threats or challenges that have emerged or are imminently promised. A danger which one state has an obligation to its own citizens and/or the international community to try to deter may have had its origin either in another state or in the use of inclusive resources. Furthermore, the state which must bear primary responsibility for deterring impending challenges need not necessarily be a state which was charged initially with the obligation to prevent the threatened effects, nor is that state or its nationals necessarily likely ultimately to be held liable for any damage which may result.

Admittedly the theory or international legal norm of state responsibility is still nebulous and underdeveloped, and rights and duties in this category are especially difficult to define. Perhaps, therefore, understanding of the problem can be facilitated through concrete illustrations of the types of circumstances that may be involved. When an injurious situation has actually emerged, there is an immediate problem of abating or minimizing the damage, but at times it may seem essential or desirable to seek to avert the risk of grave environmental danger through injunctive or other temporary relief rather than waiting for an actual threat to materialize; interstate environmental warning and notification networks are necessary to accomplish this, and some observations follow about progress to date in this regard.

ABATEMENT AND MINIMIZATION OF DAMAGE
Probably the most dramatic illustration of pollution-abatement measures is still provided by the *Torrey Canyon* catastrophe.[34] The tanker, whose deadweight tonnage of 118,285 ranked her third largest in the world at the time of her demise, was owned by a Bermuda corporation controlled by an American oil corporation, registered in and flew the flag of Liberia, and manned by an Italian crew; she was chartered by a British oil company partially owned by the British government, insured by companies in the United States and Great Britain, and claimed for salvage by a Dutch corporation. When on 18 March 1967, going at full speed, the *Torrey Canyon* struck a reef off the southwest corner of England, she was carrying 880,000 barrels of Kuwaiti crude oil. The oil started spilling out of her raked tanks almost immediately after grounding, and after three days it had covered an area over thirty-five miles long and eighteen miles wide. Carried along by the wind, the tides, and the Gulf Stream, the thick blanket of crude oil spread towards some of the best resort beaches and fishing areas in the United Kingdom. Under the circumstances, the British could hardly have been expected to sit back and wait for the rights and responsibilities to be sorted out among all the interested parties, and indeed the British government would have been sorely remiss in its responsibilities if it had failed to act to avert the danger to its own resources (not to mention those of its neighbours and inclusive resources). Consequently, after the Dutch salvors gave up their attempt to refloat the vessel, the Royal Air Force bombed her to ignite the oil remaining within the hulk so that it would burn rather than leak out into the sea. Almost two dozen large vessels and an accompanying host of smaller craft began spraying and shovelling detergent on the slick to emulsify it, but a great deal of oil still managed to pollute the waters and blacken the shores of Cornwall, of the Isle of Guernsey, and later of France.

The use of force by the RAF in the above instance was, of course, an uncharacteristic occurrence, and most oil spills occasion more routine clean-up measures. But tanker casualties themselves are, unfortunately, becoming a very familiar phenomenon. Each major world conference dealing with environmental affairs and pollution appears to have its own benchmark incident. The Stockholm Conference was marked by the Cherry Point oil spill.[35] In that incident, a Liberian tanker was unloading at the Cherry Point refinery of Atlantic Richfield Corporation located in Ferndale, Washington. She accidently spilled some twelve thousand gallons of crude oil, a good deal of which fouled about five miles of beaches in British Columbia, Canada. This spill was relatively small, but it produced major political repercussions, engendering angry Canadian newspaper articles and an emergency debate in Parliament. Meanwhile, the refinery and authorities on both sides of the border took prompt action to contain the spill and minimize the damage, and consequently 'the damage to Canadian waters and shoreline was less than might otherwise have resulted.'[36] The Caracas session of the Law of the Sea Conference had its casualty as well. When the tanker *Metula*, which flew the flag of the Netherlands, was owned by the Curacao Shipping Company, and carried Shell oil, went aground in the Straits of Magellan, losing six thousand tons of crude oil along a front of about twenty-five miles, that should have given a jolt to the Committee 3 delegates discussing the preservation of the marine environment. Fearing that the ship might split apart and thereby endanger safety of navigation, the well-being of coastal inhabitants, and the life of local marine species, the Chilean government immediately undertook emergency measures; in view of their considerable experience, the governments of Canada and the United States offered experts and technical assistance to these operations. Even the recent seventh session of UNCLOS III was celebrated by the dramatic spill of the *Amoco Cadiz* off the coast of France.[37]

It ought again to be emphasized that what is being avoided in all these cases of abatement measures is not merely impending pollution injury to waters and amenities. There is also a very clear and present danger to living resources in the case of oil spills or other casualties. As one expert has observed:

A slick kills as it goes, often doing some of its worst damage far from the scene of the disaster ... The damage a slick does as it travels is comprehensive, destroying both the very foundation of sea life, the plankton, and the highest reach of it, the birds in the air. Phytoplankton, the tiny plants responsible for photosynthesis and for the primary production of more than ninety per cent of the living material in the seas, must function in the upper levels of the oceans, where light penetrates. Fish feed on it, and the fish attract birds. Because of this cycle, all these creatures are victims of oil spills; also destroyed are the surface-mating fish eggs and fry.[38]

This excludes mention of the economic and social dependency of coastal populations on fishery resources and of national and global populations on protein resources from the sea, and so on. All this is but another illustration of the original observation here of inescapable ecological interdependencies.

Short-term measures to abate or minimize environmental harm are not, of course, exhausted by the example of oil spills. They may be concerned with activities in the air, on land, at sea, or elsewhere involving any number of possible substances and dangers. An outstanding example of the variety of factors that may be involved is provided by the Palomares incident.[39] When a US B-52 nuclear bomber collided with a jet tanker during a refuelling mission, four hydrogen bombs managed to escape. Two that fell on land ruptured, and their TNT charges exploded, scattering uranium and plutonium particles near the Spanish coastal village of Palomares and thereby causing a most grave and imminent danger to the inhabitants and the ecology of the area. The governments of the United States and Spain immediately undertook a huge effort to free the region from contamination; this even included burying some 1750 tons of mildly radioactive Spanish soil in the United States.[40] A third bomb hit the earth intact, but the fourth somehow managed to get lost. On the assumption that it lay somewhere on the deep and mountainous bottom of the Mediterranean within a 125-square miles radius, naval experts began an intense submarine search. The bomb was located after two months but tumbled over a deep underwater shelf to get lost again for nine more desperate days. At last, after eighty days, the bomb – reported to be a twenty-megaton device with an explosive force equal to twenty million tons of TNT – was retrieved in a highly specialized forty-eight-hour operation by a US Navy salvage crew. US Navy and Air Force commanders on the scene opposed any public display on security grounds, but a compromise was worked out under which certain selected Spanish officials and newsmen were allowed to view the bomb aboard the task force's flagship so as to be able to assure themselves and the public that the danger had really been averted. As far as is known, that is the first time a hydrogen bomb has ever been put on display.[41]

As all these instances illustrate, once the danger has materialized, the general nature and extent of measures that should be taken to abate it are often quite clear. In the above cases it was also easy to see who had the best means of taking them and the greatest responsibility for doing so. The latter factor is, however, not always so readily apparent. Consequently, to avoid undue haggling and critical delay, for some contingencies specific international arrangements have been made as to who is to take what action.

It is, first of all, firmly established by customary international law doctrines of self-help that any state does not have to await actual catastrophe at its borders but has a right to protect itself from impending disasters, even to the extent of employing necessary and proportional force under the circumstances.[42] And in the particular case of oil spills, the International Convention Relating to Intervention on the High Seas in Cases of Oil Pollution Casualties specifically authorizes parties to take such measures as may be necessary

to prevent, mitigate or eliminate grave and imminent danger to their coastline or related interests from pollution or threat of pollution of the sea by oil, following upon a maritime casualty or acts related to such a casualty, which may reasonably be expected to result in major harmful consequences.[43]

But conditions may arise when it is advisable or desirable that action be taken before the danger becomes 'imminent,' before it is evident which particular state is the primary target of the impending harm, or where there is reason to be greatly concerned over grave and imminent danger to shared resources; and some provisions are being made to meet these situations. Canada and the United States, for example, have adopted a Joint US-Canadian Oil and Hazardous Materials Pollution Contingency Plan for the Great Lakes Region, which fixes spheres of responsibility for the US Coast Guard and the Canadian Ministry of Transport and whose purpose is 'to provide for coordinated and integrated response to pollution incidents in the Great Lakes System by responsible federal, state, provincial and local agencies.'[44] With comparable objectives in regard to a more broadly inclusive resource, the governments of Belgium, Denmark, France, West Germany, the Netherlands, Norway, Sweden, and the United Kingdom, in the Bonn Agreement Concerning Pollution of the North Sea by Oil, have carved up that sea by longitude and latitude into zones of state responsibility '[f]or the sole purposes of this Agreement';[45] in them the parties have obligations to make assessments and take measures 'whenever the presence or the prospective presence of oil polluting the sea within the North Sea area ... presents a grave and imminent danger to the coast or related interests of one or more Contracting Parties.'[46] The International Convention for Northwest Atlantic Fisheries might also be cited in this connection, but the panels of states responsible for the subareas delineated therein are only supposed to oversee developments and then make recommendations to the governments collectively involved as to the need for measures to counteract any observed depletion of the stocks.[47] And there are undoubtedly other examples making varying arrangements in accordance with differing degrees of imminence of harm.

In addition, of course, with regard to inclusive resources in general, any state(s) or international organization(s) aware of circumstances endangering the marine or other environments may always present to the state under whose jurisdiction or control the activities concerned are being carried out, through diplomatic channels, a request for the termination or restriction of such activities and the elimination or reduction of the threat. States can take upon themselves the responsibility to back up such demands by many sorts of political, economic, and other pressures.

INJUNCTIVE AND OTHER TEMPORARY RELIEF
At times it seems essential or desirable to avert the very risk of grave danger instead of waiting until an actual threat has materialized. States may seek to accomplish this through varied channels, drawing on political and other pressures for deterrence. When the United States was contemplating 'Cannikin,' its third underground nuclear weapons test on the Aleutian Island of Amchitka, the Canadian Minister for External Affairs lodged a formal protest with the US Department of State, and Japanese officials also expressed concerned opposition.[48] These apprehensions were supplemented by several unofficial demonstrations by Canadian citizens and environmental groups, in addition to those of US environmentalists. The blast was nevertheless carried out. While no surface radiation was detected and the explosion did not cause the earthquakes that had been feared by some, it did produce high-intensity shock waves affecting large areas. In view of the magnitude of the shock waves registered in Japan, the Japanese government then sent a formal protest to the United States.

Another example of reckless or exceedingly risky behaviour, this time in connection with the use of inclusive resources, was project 'West Ford,' which determined international efforts were made to deter.[49] The US government made plans to release twenty kilograms of tiny copper 'hairs' or 'needles' in outer space to form a belt around the earth about fifteen kilometres wide and thirty kilometres deep, the objective being to test its feasibility to reflect communications signals. The prospect of such manipulation of the environment of nearby space caused international as well as national scientific groups and individual scientists very serious concern about potentially adverse effects upon radio and optical astronomy. The Soviet Union also complained that the needles might interfere with the movement of spacecraft, but that issue or pseudo-issue did not receive as much attention as the fears of the astronomers and their supporters. As a result of the many protests, a special meeting of the President's Scientific Advisory Council was called to review the project and advise whether the launching should be stopped, and PSAC deemed that it should be a safe undertaking. The first launch of West Ford consequently went ahead a month later, but the dipoles failed to disperse properly to form the belt; the final attempt was made two years later. After the success of this effort, the Soviet Union made its first formal public protest through the United Nations. A description of the subsequent propaganda manoeuvering and countermanoeuvering is not germane here. The point is that in such cases the determining factor or norm for state responsibility has to be the probability of risk rather than the wrongfulness of the conduct in and of itself. Richard Kearney, then the US member of the International Law Commission, made reference to this project on the part of his country and asked the ILC: 'Was there a question of responsibility there?'[50] As Ambassador Kearney himself gave evidence of realizing by the mere asking of the question, the answer from the perspective of the rational and progressive development of international environmental law should have been obvious.

In addition to the kind of direct effort described above, states and other actors often seek to invoke the authority of relevant courts and administrative bodies for temporary injunctive relief in aid of their causes. The hope is to obtain an official stay of the action either pending further investigation of its likely consequences or as an interim measure on the path to a final judgment disallowing it.

In the Amchitka situation, several conservation groups joined forces to seek an injunction against the tests primarily on the basis that the US Atomic Energy Commission's impact statement did not satisfy the requirements of the National Environmental Policy Act. The US District Court for the District of Columbia entered summary judgment for the AEC, but the Court of Appeals for the D.C. Circuit reversed and remanded.[51] After remand, there was a subsidiary controversy over discovery, which was itself the subject of an interlocutory appeal.[52] The District Court subsequently denied a preliminary injunction, the Court of Appeals denied the environmentalists' motion for summary reversal and a stay,[53] and the case went all the way up to the highest court in the land. The US Supreme Court, in a most rare occurrence, agreed to hear on Saturday morning a plea against the blast scheduled for that same afternoon. It rejected the last-minute appeal by a vote of four to three[54] a few minutes before the 12:30 deadline, and the five-megaton warhead went off on schedule[55] – with the dissenters still insisting that they 'would grant the injunction so that the case can be heard on the merits.'[56]

Despite this landmark case, injunctions are a fairly routine occurrence in US environmental litigation. The most notable example from the international point of view occurred in connection with the construction of the Trans-Alaska Pipeline. In an action by three US conservation groups with intervention by their Canadian counterparts and later consolidation with compatible claims of an unincorporated association of commercial fishermen, the District Court (again of D.C.) enjoined the issuance of certain permits necessary for the construction of the haul roads essential to the pipeline itself on grounds that otherwise the environmental interests 'would suffer irreparable injury.'[57] After further hearings, however, the court dissolved its preliminary injunction, denied a permanent injunction, and dismissed the complaints – which judgment was vacated and remanded, and the litigation went on.[58] There was quite a bit more litigation before the pipeline got its final go ahead directly from the U.S. Congress, but it is important that construction was at least halted for some months of further investigation while environmentalists had their day in court.

International tribunals too have been known to grant the equivalent of injunctions in environmental cases, but often with questionable effect. At an early stage of the *Fisheries Jurisdiction* cases, the International Court of Justice issued Orders Concerning Interim Measures of Protection which, among other things, provided that the sides in the two cases should 'each of them ensure that no action of any kind is taken which might aggravate or extend the dispute submitted to the Court' and also 'each of them ensure that no action is taken which might prejudice the rights of the other Party in respect of the carrying out of whatever decision on the merits the Court may render.'[59] The 1972 Orders also held that the Republic of Iceland should refrain from taking any measures to enforce its purported new fisheries regulations against ships registered in the United Kingdom or the Federal Republic of Germany outside the agreed twelve-mile fisheries zone and that that state should further refrain from applying any administrative, judicial, or other measures against such ships, their crews, or other related persons because of their having engaged in fishing activities between twelve and fifty miles off shore; for their part, the United Kingdom and Germany were directed not to take more than 170,000 and 119,000 metric tons of fish respectively from the 'Sea Area of Iceland.'[60]

A year later, having meanwhile decided that it had jurisdiction to entertain the suits[61] and being aware that negotiations had taken place between Iceland and the other states with a view towards reaching interim arrangement pending final settlement of the disputes, the ICJ issued subsequent Orders Concerning Continuance of Interim Measures of Protection.[62] Two judges lodged outright dissents, one of them placing great stress on the changed scientific and economic conditions of the fishstocks themselves, which he felt 'gave rise to questions which were serious enough to warrant inviting the Parties, before the Court took up any position on the continuance of interim measures, to furnish it with the relevant information ... as to the evolution and exploitation of the fishstocks';[63] and another, also voting against the new injunctions, questioned whether new political circumstances did not necessitate either the revocation or at least modification of the terms of the earlier orders on grounds that they were obviously ineffective:

The reason is that, as no one can be unaware, there have been numerous clashes in the disputed fishery-zone between Icelandic coastguard vessels and trawlers flying the British or Federal German flag. Some of these incidents, such as collision between two vessels or the firing of shells by Icelandic coastguard vessels, were in my view grave enough to warrant the exercise by the Court of its right to modify the terms of its original decision.

Furthermore, these incidents ... constitute so many flagrant violations on either side of the operative part of the Orders of 17 August 1972. The measures should therefore be reviewed and others indicated concerning *inter alia* the presence of warships.[64]

In any case, Iceland continued to disregard the injunctions against it, although a few months after the second set of orders, Iceland and Great Britain reached an 'Interim Agreement in the Fisheries Dispute'[65] (but such was not the result of similar discussions with the Federal Republic). Finally, as has been recorded elsewhere here, by ten votes to four, the World Court decided against Iceland on the merits but held that all parties in interest were under mutual obligations to negotiate in good faith for the equitable solution of their differences.[66]

A more unencouraging instance of what were effectively international preliminary injunctions in an environmental controversy occurred in the *Nuclear Tests* cases. In the course of the challenges by Australia and New Zealand to French atmospheric weapons tests over the South Pacific, the International Court of Justice again felt called upon to issue Orders Concerning Interim Measures of Protection. By eight votes to six, the Court instructed France to 'avoid nuclear tests causing the deposit of radioactive fall-out' over Australia and New Zealand pending final decision in its proceedings.[67] The orders came down on 22 June 1973, and less than one month later, on 21 July, France exploded another device over its Pacific atoll of Mururoa; the French government also went ahead with another series of tests less than a year later. Subsequently, however, it will be recalled that various French officials issued unilateral statements of intention to cease atmospheric explosions and pass on to the stage of underground tests after the 1974 blasts, thus allaying the fears of the two applicants as regards radioactive danger to their people and resources. In the light of these declarations, as has been discussed earlier in this chapter, the ICJ chose not to make a declaratory judgment on the legality of atmospheric nuclear testing or to render any other judgment in the case. It may be, however, that ultimate French policy and actions were influenced by the very fact of the juridical proceedings themselves.

WARNING AND NOTIFICATION

During the preparatory processes of the Stockholm Conference, Principles 21 and 22 on state responsibility and liability were accompanied by a third principle relating to the duty to provide proper warning to other states. Although it has already been quoted elsewhere, 'draft Principle 20' as originally proposed by the Working Group on the Declaration is also worth repeating in the present context:

Relevant information must be supplied by States on activities or developments within their jurisdiction or under their control whenever they believe, or have reason to believe, that such information is needed to avoid the risk of significant adverse effects on the environment in areas beyond their national jurisdiction.[68]

Some delegations contended, however, that any such consultation responsibilities were inappropriate for inclusion in the Declaration on directly opposite grounds: it was contended, on the one hand, that they 'were inherent in the obligations undertaken by Member States in the Charter of the United Nations,' and, on the other, that they 'were an extension of these obligations which would be outside the scope of a declaratory and inspirational instrument.'[69] There was much further debate and a whole spectrum of opinion as to the proper nature and content of an acceptable compromise.

No solution was reached either before or at the Conference itself. In Stockholm the recognition of a duty to warn was effectively blocked by the Brazilian delegation. Brazil was at the time undertaking feasibility studies for a giant hydroelectric installation on the Parana River, which eventually flows into Argentina, becoming the La Plata. Argentina feared that such an alteration in the flow of the river might cause floods, droughts, water pollution, and other injury to Argentine environmental interests, and that government therefore called upon its neighbour to enter into consultations before going ahead with the plans. Accordingly, Argentina proposed at the Conference that the principle be strengthened by adding: 'This information must also be supplied at the request of any of the Parties concerned, within appropriate time, and with such data as may be available and as would enable the above-mentioned Parties to inform and judge by themselves of the nature and probable effects of such activities;'[70] Brazil, on the contrary, wanted expressly to limit it: 'No State is obliged to supply information under conditions that, in its founded judgment, may jeopardize its national security, economic development or its national efforts to improve environment.'[71] The Conference finally avoided deciding the question by referring it to the General Assembly the next fall, in the hope that a consensus might have emerged by that time.

In the General Assembly a few months later, Brazil took the lead in coming forth with something of a conciliatory proposal. The substantially weakened resolution, which was co-sponsored by a large number of developing countries and a few developed countries, recognized

that co-operation between States in the field of the environment, including co-operation towards the implementation of principles 21 and 22 of the Declaration of the United Nations Conference on the Human Environment, will be effectively achieved if official and public knowledge is provided of the technical data relating to the work to be carried out by States within their national jurisdiction, with a view to avoiding significant harm that may occur in the environment of the adjacent area.[72]

It then further stipulated that such 'technical data' should be given and received in the 'best spirit of co-operation and good-neighbourliness' and without its being used 'to delay or impede the programmes and projects of exploration, exploitation and development of the natural resources of the States in whose territories such programmes and projects are carried out.'[73] It is clearly this last caveat, intended to stress the paramountcy of economic and social development, that won the support of the 'Group of 77' for General Assembly Resolution 2995 (XXVII). While on balance according it their support, however, some countries were worried about the implications that might be drawn from the feeble character of 2995 as regards the overall concept of state environmental responsibility. They therefore promoted concurrent passage of General Assembly Resolution 2996 (XXVII), which states flatly that 'no resolution adopted at the twenty-seventh session of the General Assembly can affect principles 21 and 22 of the Declaration of the United Nations Conference on the Human Environment.'[74]

Governments soon began to realize, nevertheless, that the exhortation to provide 'technical data' and no more in resolution 2995 was simply not enough from the point of view of environmental protection. Consequently, the next year the General Assembly passed a new and stronger resolution on 'Co-operation in the Field of the Environment Concerning Natural Resources Shared by Two or More States.' General Assembly Resolution 3129 (XXVIII) specifically considers that 'it is necessary to ensure effective co-operation between countries through the establishment of adequate international standards for the conservation and harmonious exploitation of natural resources common to two or more States in the context of the normal relations existing between them'; and it considers further that 'co-operation between countries sharing such natural resources and interested in their exploitation must be developed on the basis of a system of information and prior consultation within the framework of the normal relations existing between them.'[75] It therefore requests the Governing Council of the United Nations Environment Programme to report on measures adopted for the implementation of these two paragraphs and solicits UNEP and member states to take them fully into account.

The second session of the UNEP Governing Council accordingly requested the Executive Director to prepare a study and make proposals along these lines.[76] After consulting widely with governments and international organizations, the Executive Director proposed in his report, among other things, that there should be consideration of a draft code of conduct setting forth general principles and guidelines for the conduct of states in the conservation and harmonious exploitation of shared natural resources.[77] Accordingly the third Governing Council authorized him to establish an intergovernmental working group to begin the drafting; this decision was not reached, however, without a fair amount of doubt and hesitation, with the roll call vote being twenty-eight votes to one with twenty abstentions (the single negative vote being cast by Brazil).[78] However that may be, the Executive Director did establish the working group, which proceeded to draft a comprehensive set of principles concerning shared natural resources.[79]

In some cases and with regard to certain problems, states have very recently begun on a multilateral and bilateral basis to try to elaborate upon and implement these responsibilities for information exchange and prior consultation. The 1974 Principles Concerning Transfrontier Pollution of the Organization for Economic Co-operation and Development specify both that countries concerned should exchange 'all relevant scientific information and data on transfrontier pollution' (an advance over merely 'technical data' or plain 'information' in the previous formulations) and also that they should 'promptly warn other potentially affected countries of any substances which may cause any sudden increase in the level of pollution in areas outside the country of origin of pollution' (which may be taken to imply a responsibility to analyse available information and present it in a readily assimilable form).[80] The 1975 Canada-US Agreement on the Exchange of Information on Weather Modification Activities – to return to another example discussed in chapter 3 – not only tries to elaborate upon what type of information is to be communicated, in what form, and by and to whom, but also commits the responsible agencies in each country to 'consult with a view to developing compatible reporting formats, and to improving procedures for the exchange of information.'[81] In keeping with General Assembly Resolution 3129, the parties additionally 'agree to consult ... regarding particular weather modification activities of mutual interest,' except in extreme emergencies,[82] but the manner and procedures for incorporating such prior consultations into the conduct of their normal relations is not spelled out.

Finally, at the Law of the Sea Conference, the international community has been doing some work clarifying and expanding upon evaluation and notification requirements. A draft article has been adopted which obligates states having reasonable grounds for expecting that planned activities under their jurisdiction or control may cause substantial pollution of the marine environment both to 'assess the potential effects of these activities' and to 'communicate reports' of them. At first its value as a means for the transfer of information and the issuance of environmental warning may seem to be somewhat wanting, since states are merely to communicate the reports 'to the competent international or regional organizations, which should make them available to all States.'[83] Yet this draft article on environmental assessment must be interpreted in the light of another draft article on global and regional co-operation, which provides that a state which becomes aware of cases in which the marine environment is in imminent danger of being damaged or has been damaged by pollution 'shall immediately notify other States it deems likely to be affected by such damage, as well as the competent international organizations, global or regional.'[84] Taken together, the combination of responsibilities is not inconsequential.

In short, slowly and tentatively, there is developing a legal norm of environmental warning and notification the violation of which may entail international responsibility on the part of states.

Reparation or compensation for environmental injury

When environmental deprivations have not been prevented or potential hazards have not been effectively deterred, there then arises the problem of reparation of the harm and compensation for the damages that have occurred. It is in connection with this subgoal of legal sanctioning processes that the reference to concern over 'international law regarding liability and compensation for the victims of pollution and other environmental damage' in Principle 22 of the Stockholm Declaration becomes of direct concern here.[85] As has been explained by other writers at length, there are actually three relevant subsidiary aspects to the problem of what is to be done after the fact of injury: 'restoration' of relationships between or among the parties involved, 'rehabilitation' or immediate reparation of the values destroyed, and 'reconstruction' or long-term avoidance of unauthorized manipulation of the basic value relations at issue.[86] Since international environmental law is still in such a relatively primitive stage of development, however, it is not often possible to distinguish between these aspects in considering existing precedents.

The basic question here is the standard of liability to be applied. It may obviously make a great deal of difference in the respective rights and value positions of the parties concerned if the basis of liability is fault or negligence, on the one hand, or there is strict or absolute liability, on the other. And it is highly important to the protection and preservation of the environment that liability follow upon mere demonstration of causality of injury, rather than only upon proof of intention to harm or some other wrongful behaviour. A second fundamental and related question is who is to be held liable. Assuming that liability attaches to the type of conduct involved, it is clear that a state is liable for damage attributable or imputable to it. When damage has been caused by its nationals or some activity under its jurisdiction or control, however, a state may choose in the first instance to provide appropriate recourse directly against the natural or juridical person (s) involved; but following exhaustion of any such local remedies, the state that is itself injured or is the national state of a damaged party still has the right to present a claim to the responsible state, and the latter state would then be answerable for any compensation found to be due (although it would probably, of course, then seek to extract payment from the particular relevant actor or actors). Under some circumstances, nevertheless, reciprocally or otherwise, states may wish to assume liability in the first instance for certain activities of their nationals. Further complications can arise in connection with both of these questions. For example, designation of liability and assessment of the nature and extent of damages are likely to be much more difficult when injury has been done to inclusive interests rather than exclusive interests. And on the matter of the respository of the liability, situations are not at all unforeseeable where there will be joint liability and perhaps indemnity from one responsible state or other party to another.[87]

International law on many of these issues is, to repeat, not at all clear. Certain consistent and generalizable patterns of reparation and compensation requirements can, nevertheless, be observed in the contemporary development of international environmental law. The next section discusses the standard of strict or absolute liability both as it is emerging under general circumstances and as it has been applied to certain ultrahazardous activities; then, since the possibility and feasibility of insurance underlies so much of the policy consideration in this area, a commentary on the nature and functions of various environmentally related compensation funds follows.

EMERGENCE OF STRICT LIABILITY
FOR ENVIRONMENTAL INJURY

'Strict liability' is liability without fault, and it may be said to exist when compensation is due from one actor to another for injuries caused despite compliance with any particular standards of care. A number of variations on the theme of strict liability have evolved in common law jurisdictions (for example, nuisance, ultrahazardous activities, trespass, and borderline doctrines such as *res ipsa loquitur*), but for present purposes a general comprehension of the generic term should suffice. It cannot be said that there are no defences to strict liability since, depending on the degree of strictness involved, it may be subject to the classic exonerations for tortious acts: *force majeure*, acts of God, or interventions of other third parties. Consequently, while some writers use the terms 'strict' and 'absolute' liability interchangeably, others prefer to reserve the latter for conditions under which very few or no exculpations apply; usage here will attempt to follow that general distinction.[88] But all of these are just words, none of which is very precise, and it must be recognized that what goes on in any given case is a balancing of multiple and multidimensional interests rather than merely pinpointing along some non-existent linear theoretical projection.

One professor has maintained that 'the *Trail Smelter*, *Corfu Channel* and *Lac Lanoux* cases clearly point to the emergence of strict liability as a principle of public international law.'[89] Others who have analysed these very few precedents in the field of international environmental law usually tend to agree with him that there is an evolving norm of strict liability for environmental injury modelled on the century-old rule adumbrated in the famous case of *Rylands* v *Fletcher*.[90] In that case, the defendants, who were proprietors of a mill, had built a large reservoir on their own land for their own business purposes. It was perfectly lawful for them to do this, and they employed for the project a competent engineer and competent contractors. Unfortunately, however, due to the unknown and unsuspected presence of an old and abandoned mine shaft, water leaked out of the reservoir and flooded the tunnels in the mine operated by the plaintiff on his adjoining property. The plaintiff therefore sued to recover the damages caused by the flooding of his mine, and the English House of Lords found in his favour. The theory of the case was that the mill owners had put their land to a 'nonnatural use' by collecting on it an unusual amount of water and that they were consequently liable for damage caused to someone else's property; as the Lord Chancellor observed, 'that which the Defendants were doing they were doing at their own peril.'[91]

Parallels in the *Trail Smelter*[92] situation are not difficult to find. That controversy, it will be remembered, involved damage occurring in the territory of the United States and alleged to be caused by an agency situated in Canada. The Consolidated Mining and Smelting Co. of Canada Ltd was operating a smelter at Trail, British Columbia, which was one of the largest and best-equipped such plants on the North American continent. Due at least in part to certain characteristics of river and air currents in the valley shared by the two countries, the fumes were claimed to be causing air pollution and damage to crops in the increasingly populated farming areas around Northport, a town in the State of Washington. The arbitral tribunal set up to resolve the matter found that the Dominion of Canada was responsible at international law for the conduct of the mining company in Canadian territory, that the damage south of the border was indeed caused by the operation of the Canadian smelter, and consequently that indemnity was due from Canada to the United States in compensation for the injury. It based its decision on the much-quoted observation that 'no State has the right to use or permit the use of its territory in such a manner as to cause injury by fumes in or to the territory of another or the properties or persons therein ... '[93]

An even clearer illustration of the application of strict liability in the context of environmental injury is the *Gut Dam*[94] arbitration between the same two countries. The facts have been summarized in an earlier chapter but bear repeating here in less skeletal form. In 1874 the Canadian Chief Engineer of Public Works proposed to his government that it construct a dam between Adams Island in Canadian territory and Les Galops, in the United States, for the purpose of improving navigation in the St Lawrence River. After many investigations and reports and formal approval by an act of the US Congress, the Canadian government proceeded to construct the dam in 1903; experience soon demonstrated that it was too low to serve the desired ends and, again with explicit US permission, Canada increased the height of the dam a year later. Between 1904 and 1951 several man-made changes affected the flow of water in the Great Lakes – St Lawrence River basin, and, while the dam itself was not altered in any way, the level of the water in the river and nearby Lake Ontario increased. In 1951–2 the level of waters reached unprecedented heights which, in combination with storms and other natural phenomena, resulted in extensive flooding and erosion damage on both the north and south shores of all the lakes. In 1953, the government of Canada removed its dam as part of the construction of the St Lawrence Seaway, but the problem of US claims for damages allegedly resulting from the presence of Gut Dam still festered for some years. The Lake Ontario Claims Tribunal set up to resolve these matters, after initially determining that Canada had an obligation to all citizens of the United States and not just the owner of Les Galops regarding the construction of the dam and that such responsibility was not limited in time to some initial testing period, observed that 'the only issues which remain for its consideration are the questions of whether Gut Dam caused the damage for which claims have been filed and the *quantum* of such damages.'[95] The arbitral tribunal was, in other words, clearly adopting a standard of strict liability, since it was not interested in hearing any arguments for or against fault or negligence in planning and construction or whether Canada knew or ought to have known what injuries might result. Following upon this holding by the Claims Tribunal, the two governments concerned reached a negotiated settlement of a lump-sum payment from Canada to the United States 'in full and final satisfaction of all claims of United States nationals for alleged damage caused by Gut Dam,'[96] which was then approved by the Tribunal.[97]

It is worth mentioning in passing that these two neighbours have been at it again – this round with the parties reversed. The controversy this time concerns the Garrison Diversion, a planned $400 million irrigation project. In 1965, the US Congress authorized the project, designed to divert Missouri River water into central and eastern North Dakota. Officials in Manitoba became alarmed that the resultant drainage and wastewaters would degrade the water quality of the Souris, Assiniboine, and Red Rivers as well as Lakes Manitoba and Winnipeg and would cause other problems.[98] The Canadian concerns, crystallized in an aide-memoire, prompted discussions between the United States and Canada in 1970.[99] Subsequently, the government of Canada invoked the provision of the 1909 Boundary Waters Treaty that such waters 'shall not be polluted on either side to the injury of health or property on the other'[100] and referred the matter to the International Joint Commission. The IJC established the Garrison Diversion Study Board to undertake a technical investigation. As a result of its studies, the Commission recommended in August 1977 that the portion of the proposed Garrison Diversion which would affect waters flowing into Canada not be built at that time and outlined its recommendations of conditions under which it believed the portion might later proceed.[101] The report is now before the two governments for their consideration. This problem of the Souris and other rivers is only one of several environmental disputes simmering in recent years in North America,[102] and environmental problems are not necessarily any more solicitous of national boundaries elsewhere.

Turning now to the other two cases cited at the outset of this section, *Corfu Channel*[103] and *Lac Lanoux*,[104] the existence of strict liability is not difficult to maintain. The former case involved a finding by the International Court of Justice that the People's Republic of Albania was liable for the consequences when British warships struck mines in the Albanian waters of the Corfu Channel. More exactly, the conclusion of the Court was 'that Albania is responsible under international law for the explosions which occurred ... and for the damage and loss of human life which resulted from them, and that there is a duty upon Albania to pay compensation to the United Kingdom.'[105] The liability stemmed directly from the presence of the mines and the failure to warn the approaching vessels, with no proof required of any malevolence, neglect, or other wrongfulness on the part of Albania. By contrast, the *Lac Lanoux* arbitration was a suit by Spain to block before it was undertaken a hydroelectric project by France using the waters of the lake. It was decided that the proposed project would not be in violation of France's obligation under treaties with its neighbour or under general international law, since it was found to represent a reasonable utilization of the water resources that should not prove injurious to Spanish interests. The tribunal nevertheless added that, if the works did in fact cause pollution or other actual damage, 'Spain could then have claimed that her rights had been impaired.'[106]

Besides these few juridical requirements, there have been certain other instances of voluntary compensation for environmental injury by governments or other actors without proof of fault, but their precedential value is questionable and limited. A much-cited example is the *ex gratia* payments by the United States government to the Japanese government as compensation for Japanese nationals who sustained personal and property damage as a result of the nuclear tests in the Marshall Islands in 1954. Although the tests themselves may be considered lawful measures for security, due to a series of miscalculations or for some other reason, a number of Marshallese, Japanese, and Americans were injured by the test of 1 March of that year, and the series as a whole somewhat disrupted activities of the Japanese fishing industry.[107] The United States tendered $2 million to Japan 'for purposes of compensation for the injuries or damages sustained' and 'in full settlement of any and all claims against the United States or its agents, nationals or juridical entities' as a result of the tests, but did so 'without reference to the question of legal liability.'[108] To take a more recent example on the part of a private actor, the Atlantic Richfield Company, which operated the refinery at Ferndale, Washington, that was the site of the 1972 Cherry Point oil spill, paid an initial clean-up bill of $19,000 submitted by the municipality of Surrey for its activities. ARCO later agreed to pay another $11,606.50 to be transmitted by the United States to the Canadian government for its costs incurred in connection with clean-up operations, but would not consent to provide reimbursement for an additional item of $60 designated 'bird loss (30 birds at $2 a bird).' Again this was done, however, 'without admitting any liability in the matter and without prejudice to its rights and legal position.'[109]

IMPOSITION OF ABSOLUTE LIABILITY
FOR ULTRAHAZARDOUS ACTIVITIES

Many systems of municipal law contain rules creating 'absolute' or exceedingly strict liability for failure to control operations which necessarily create a serious or unusual risk of harm to others. Such designations are, to repeat, only words, and it is essential to look at the varied factors and policies behind them. These rules are based, at least in part, upon principles of loss distribution and liability imposed upon the effective (insured or self-insured) defendant. Ian Brownlie reports that it is the general opinion that international law at present lacks such a doctrine, although Wilfred Jenks has proposed that the law be developed on the basis of a Declaration of Legal Principles Governing Ultrahazardous Activities Generally, which would be adopted by the UN General Assembly.[110] But he also acknowledges that caution is required in accepting the statement that existing law lacks such a principle, 'because the operation of the normal principles of state responsibility may create liability for a great variety of dangerous activities on state territory or emanating from it.'[111] And in every event, absolute liability has at least been recognized in several multilateral conventions as regards certain exceptionally risky or hazardous activities. Of particular interest for present purposes is the increasing imposition of absolute liability in respect of nuclear installations and the operation of nuclear ships, damage caused by space objects, and certain types of oil pollution incidents.

Dealing with a relatively new and obviously dangerous enterprise, the drafters of the 1962 Brussels Convention on the Liability of Operators of Nuclear Ships were quite straightforward: 'The operator of a nuclear ship shall be absolutely liable for any nuclear damage upon proof that such damage has been caused by a nuclear incident involving the nuclear fuel of, or radioactive products or waste produced in, such a ship.'[112] The only narrow exculpation allowed is for intentional wrong on the part of the injured party: 'If the operator proves that the nuclear damage resulted wholly or partially from an act or omission done with intent to cause damage by the individual who suffered the damage, the competent courts may exonerate the operator wholly or partially from his liability to such individual.'[113] The operator may have an independent right of recourse against some other party, but this basic liability is certainly strict enough to deserve its appellation of 'absolute'; the extent of the liability in respect of any one nuclear incident was, however, limited to approximately $100 million.[114]

The 1963 Vienna Convention on Civil Liability for Nuclear Damage was equally succinct: 'The liability of the operator [of a nuclear installation] for nuclear damage under this Convention shall be absolute'[115] – although its exculpatory clause is slightly broader, allowing the competent court if its law so provides to relieve the operator wholly or partly if he proves that the nuclear damage resulted 'either from the gross negligence of the person suffering the damage or from an act or omission of such person done with intent to cause damage';[116] the Vienna Convention allows states to set limits of not less than $5 million for any one nuclear incident.[117] The above are but two of four early conventions in the field of nuclear liability,[118] the others being the 1960 Paris Convention on Third Party Liability in the Field of Nuclear Energy[119] and the 1963 Convention Supplementary to this Paris (OECD) Convention.[120] As in the present context we are interested in the strictness of the standard of liability specified, the precise mechanics of these agreements need not be explored here. It is nevertheless worth noting that, in keeping with their primary concern being the risk inherent in the activities themselves involved, these international conventions treat all nuclear operators – whether government agencies or private corporations – on a similar basis. Finally, as concerns maritime carriage of nuclear material, in order to ensure that the operator of a nuclear installation will be exclusively liable for damage caused by a nuclear incident, the Vienna and Paris Conventions have been complemented by a fifth agreement; the 1971 IMCO Convention Relating to Civil Liability in the Field of Maritime Carriage of Nuclear Material exonerates any and all other persons.[121]

Another area for which there has been designation of liability so strict as to verge on being absolute is the exploration and exploitation of outer space. The various expressions of world community expectations in this realm have already been recorded. The essential point to note is that, while the 1967 Outer Space Treaty states only most generally and summarily that a launching state 'shall bear international responsibility' and 'is internationally liable for damage to another State Party to the Treaty or to its natural or juridical persons by such object or its component parts on the Earth, in air space or in outer space, including the moon and other celestial bodies,'[122] there has been some further clarification of the nature of the liability. The 1971 Convention on International Liability for Damage Caused by Space Objects expressly provides that a launching state 'shall be absolutely liable to pay compensation for damage caused by its space object on the surface of the earth or to aircraft in flight.'[123] There may be exoneration from this absolute liability in accordance with the broader standard, that is, if the state responsible for the launch establishes that 'the damage has resulted either wholly or partially from gross negligence or from an act or omission done with intent to cause damage on the part of a claimant State or of natural or juridical persons it represents,' but no exoneration whatsoever is to be granted in cases where the damage has resulted from activities conducted by a launching state 'which are not in conformity with international law including, in particular, the Charter of the United Nations and the Treaty on Principles Governing the Activities of States in the Exploration and Use of Outer Space, including the Moon and Other Celestial Bodies.'[124] By contrast, where two ultrahazardous activities or in this instance space expeditions clash, in other words where one space object causes injury to another or to persons or property on board, then a fault standard is to be reinstated and the launching state 'shall be liable only if the damage is due to its fault or the fault of persons for whom it is responsible.'[125]

Finally, as the potential hazards of oil spills and other catastrophes are increasing exponentially, more and more strict liability may be finding a place in the legislative regime of the law of the sea as well. It has already been mentioned in passing that liability under the Canadian Arctic Waters Pollution Prevention Act 'is absolute and does not depend upon proof of fault or negligence' (with certain relatively minor exceptions).[126] There is, in addition, some indication of movement multilaterally towards such a standard. During the negotiations of the 1969 IMCO International Convention on Civil Liability for Oil Pollution Damage,[127] there was considerable debate as to whether the applicable standard should be fault or strict liability.[128] Even after the *Torrey Canyon* disaster, there was substantial although diminishing support for a fault basis. But after much deliberation, by the time of the Brussels Conference which concluded this IMCO 'Private Law' Convention, the tide of international opinion had clearly shifted to favour a rather strict standard. The resultant liability provision is worth quoting at length, to give an overall idea of how such requirements are put together:

1 / Except as provided for in paragraphs 2 and 3 of this Article, the owner of a ship at the time of an incident, or where the incident consists of a series of occurrences at the time of the first such occurrence, shall be liable for any pollution damage caused by oil which has escaped or been discharged from the ship as a result of the incident.

2 / No liability for pollution damage shall attach to the owner if he proves that the damage:

a / resulted from an act of war, hostilities, civil war, insurrection or a natural phenomenon of an exceptional, inevitable and irresistible character, or

b / was wholly caused by an act or omission done with intent to cause damage by a third party, or

c / was wholly caused by the negligence or other wrongful act of any Government or other authority responsible for the maintenance of lights or other navigational aids in the exercise of that function.

3 / If the owner proves that the pollution damage resulted wholly or partially either

from an act or omission done with intent to cause damage by the person who suffered the damage or from the negligence of that person, the owner may be exonerated wholly or partially from his liability to such person.[129]

This is obviously not 'absolute' liability, being not nearly as strict as the nuclear or space formulations; but, even without the 'wholly' restriction in paragraph 2, it could still be argued to be a bit stricter than most traditionally allowable defences under international maritime law. The trend of development by which it was reached is, of course, also significant. Moreover, whatever its precise categorization, this convention is also notable for being backed up by supplementary arrangements setting up compensation funds for oil pollution damage.

INTERNATIONAL ENVIRONMENTAL COMPENSATION FUNDS
The Brussels Conference which adopted the International Convention on Civil Liability for Oil Pollution Damage also passed a Resolution on Establishment of an International Compensation Fund for Oil Pollution Damage, which recommended early establishment of such an insurance fund founded on two basic principles:

1 / Victims should be fully and adequately compensated under a system based upon the principle of strict liability.
2 / The fund should in principle relieve the shipowner of the additional financial burden imposed by the present ['Private Law'] Convention.[130]

Accordingly, at another IMCO Conference in Brussels two years later, a new agreement was concluded towards these ends. Although its preamble states that the 1971 International Convention on the Establishment of an International Fund for Compensation for Oil Pollution Damage[131] is to be 'supplementary to' the earlier provisions, it is only somewhat so. On the one hand, it did raise the amount of compensation available from the then $14 million limit of the IMCO 'Private Law' Convention to about $34 million (now around $17 and $36 million denominated in terms of 'Special Drawing Rights').[132] On the other hand, however, it still did not expand coverage to pollutants other than oil, nor does it encompass damage outside the 'territory including territorial sea' of the parties.[133] As far as procedure is concerned, when the Fund Convention comes into force on 16 October 1978 and its Oil Pollution Fund is created, contributions in respect of each state party are to be paid in by the petroleum industry on the basis of tons of oil received in ports or terminal installations.[134] Actions for compensation or indemnity will be able to be brought against the

Fund in the national courts of the state or states in which the pollution damage has been caused or measures have been taken to prevent or minimize damage.[135]

At the time of this writing, the Oil Pollution Liability Convention has only recently come into effect, and the supplementary Fund Convention enters into force by the end of the year. Meanwhile, however, the provisions of the Tanker Owners Voluntary Agreement concerning Liability for Oil Pollution[136] and the Contract Regarding an Interim Supplement to Tanker Liability for Oil Pollution,[137] known as TOVALOP and CRISTAL, have been applicable.[138] TOVALOP, as its name indicates, is a contract among the tanker owners to the effect that, '[i]f a discharge of Oil occurs from a Participating Tanker through the negligence of that Tanker (and regardless of the degree of fault)' and in addition 'if the Oil causes Damage by Pollution to Coast Lines within the jurisdiction of a Government or creates a grave and imminent danger of Damage by Pollution thereto,' then the tanker owner is obliged to remove the oil or to reimburse the government of the coastal state for the removal costs incurred.[139] Maximum liability in any one incident is limited to the lower of $100 (later $125) per gross registered ton or $10 million.[140] Although it does offer something by way of reparation, TOVALOP has certain readily apparent disadvantages, such as requiring proof of negligence, running in favour only of governments, and applying only to oil removal costs and not to other measures of deterrence of or compensation for damage.

The other agreement, CRISTAL, is an oil company attempt to supplement the compensation provisions of the 'Private Law' Convention and TOVALOP pending entry into force of the Fund Convention. It is a contract among the companies to create a fund out of which public or private persons can be compensated for pollution damage up to $30 million per incident.[141] The Oil Companies Institute for Marine Pollution Compensation Ltd, a Bermuda entity organized to administer the fund of CRISTAL, is liable only in cases where liability arises under the terms of TOVALOP or the Liability Convention.[142] As this is being written, both TOVALOP and CRISTAL are in the course of revision. Their limits of liability are expected to be increased, and many of the defects discussed here should be eliminated.[143]

The oil companies are not concerned with their liability solely as regards tanker ownership and operation. In view of the growing impetus towards exploration and exploitation of the resources of the continental shelf and the seabed, the same basic group of corporations felt called upon in late 1974 to conclude an Offshore Pollution Liability Agreement (OPOL).[144] In many respects, OPOL functions for offshore oil facilities as the TOVOLOP and CRISTAL fund and insurance schemes do with respect to owners and operators of oil tankers. The new contract is not restricted to negligence and does provide for compensation to private persons. Its main operative paragraph on remedial measures, reimbursement, and compensation for claims states:

If a Discharge of Oil occurs from a Designated Offshore Facility, and if, as a result, any State or States take Remedial Measures and/or any Person sustains Pollution Damage, then the Party hereto who was the Operator of said Designated Offshore Facility at the time of the Discharge of Oil shall reimburse the cost of said Remedial Measures and pay compensation for said Pollution Damage up to an overall maximum of U.S. $16,000,000 per Incident ... [145]

In the same clause, exceptions are allowed for damage which resulted from acts of war or other hostilities, was wholly caused by an act or omission done with intent to cause damage by a third party, was wholly caused by the negligence or other wrongful act of any state or other licensing authority, or resulted from an act or omission done with intent to cause damage or from negligence of the claimant – that is, fairly standard strict liability exonerations. The signatories formed under the laws of England the Offshore Pollution Liability Association Ltd for the purpose of administering this contract and certain other functions.[146] Furthermore, there is also now the 1976 Convention on Civil Liability for Oil Pollution Damage Resulting from Exploration and Exploitation of Seabed Mineral Resources, which stands in very much the same relation to OPOL as the 1969 and 1971 Civil Liability and Fund Conventions do to TOVOLOP and CRISTAL; in the sea-bed mining case, however, the convention is limited in its applicability to the Western European region and differs significantly in several of its terms from OPOL.[147]

The above examples all deal with internationally and transnationally organized environmental compensation funds. Funds created in pursuance of national legislation may also have international significance, of which a noteworthy example is that established under the US Trans-Alaska Pipeline Authorization Act of 1973.[148] After providing for strict liability for activities in connection with the pipeline right of way on the part of the holder,[149] the Act goes on to create a compensation fund to pay for related incidents. The liability provision is as follows:

Notwithstanding the provisions of any other law, if oil that has been transported through the trans-Alaska pipeline is loaded on a vessel at the terminal facilities of the pipeline, the owner and operator of the vessel (jointly and severally) and the Trans-Alaska Pipeline Liability Fund established by this subsection, shall be strictly liable without regard to fault in accordance with the provisions of this subsection for all damages, including clean-up costs, sustained by any person or entity, public or private, including residents of Canada, as a result of discharges of oil from such vessel.[150]

Exoneration is allowed for damages caused by act of war or the negligence of the United States or other governmental agency and for the negligence of the damaged party, and liability is limited under this subsection to $100 million for any one incident.[151] The Fund itself is to be a non-profit corporate entity that may sue and be sued in its own name, with its resources to be maintained at the level of $100 million by imposition of a fee of five cents per barrel by the pipeline operator against the owner of the oil.[152] Aside from the immediate and obvious interests of Canada and Canadian citizens in this piece of US legislation,[153] it is a matter generally worthy of international concern whether or not provision is made sufficiently to take into account possible environmental costs and ways of meeting them in such a large-scale undertaking as this.

A NOTE ON PRIVATE RIGHTS OF
ACTION UNDER NATIONAL AND LOCAL LAWS
Since the present focus of inquiry is on 'state responsibility' in particular within the broader context of 'world public order,' there has been little discussion here of private rights of action in domestic arenas. This should by no means be taken to imply that the possibility of private litigation or other action is not of great significance. Sometimes it is found preferable to international solutions. In *Michie* v *Great Lakes Steel Division, National Steel Corp.*,[154] for example, plaintiffs who are residents of the Windsor area in Canada and regular recipients of effluents from the Detroit industrial complex in the United States, chose to institute private actions against three industrial corporations and seek remedies for their air pollution in US courts under municipal law rather than wait for a long and tedious international claims process as in *Trail Smelter* or *Gut Dam*.[155] And the new Nordic Environmental Protection Convention,[156] of course, envisions such suits happening on a regular and usual basis among citizens and entities of the four countries involved. Furthermore, in some cases private arrangements may be made to provide rapid compensation without the need for juridical intervention of any type, as the US government undertook to do in regard to 597 claims filed in connection with the Palomares incident[157] and as it is reported ARCO did privately by opening up a claims office to handle claims resulting from the Cherry Point oil spill.[158]

Frequently such private litigation under national and local laws will be capable of resolving the controversy satisfactorily and expeditiously. Often, nevertheless, the juridical tangle likely to result without benefit of additional international liability and compensation arrangements boggles the mind. There was, for example, a good deal of private litigation – some of it effectively resolved – in the wake of the *Torrey Canyon* disaster.[159] Yet in most such situations, with all the parties and countries involved with different applicable fault and limitation provisions and other conflicting legal requirements, truly prompt and adequate compensation seems virtually unimaginable without prior general agreement on a standard of strict liability for some state, company, or other international actor involved imposed in accordance with certain set rules and regulations. And it is only through such agreement that the parties in interest can know reliably beforehand the possible consequences of their activities so as to be able to make arrangements for insurance or other cost-spreading devices.

In short, although this is not the place to discuss comparative national substantive legislation or procedural requirements in environmental cases, it must be noted that municipal law, too, can be critically important. It is unrealistic to try to make any sharp distinctions between international and national legal responsibilities, as these involve multiple interacting and interreactive processes of authoritative decision.

7

International environmental
dispute settlement

As has been emphasized throughout, the earth-space environment is an ecological unity both in the basic scientific sense and in the interdependencies of the social processes by which mankind uses it. Viewed ecologically, however, international politics must be seen as a system of relationships among multiple and interpenetrating earth-related communities that share with one another an increasingly crowded planet offering finite, exhaustible, and destructible quantities of basic essentials for physical sustenance and a life of dignity and well-being. Thus, while all states and all peoples share common interests in the protection and preservation of the earth-space environment, they have concomitant competing interests in securing certain of its limited benefits. As a result, although a primary goal of environmental world public order is the prevention of environmental disputes, some controversies are undoubtedly inevitable. In contrast to the last chapter, which looked towards a new and largely unexplored area of intenational law, this chapter analyses traditional international dispute settlement mechanisms and practices as applicable in the environmental context. Such a review will, it is hoped, demonstrate that what is required in this connection is not new organizational arrangements *per se*, but rather a reorientation of values and a reordering of priorities within the existing processes in accordance with emerging ecological imperatives.

Two distinct forms of jurisdictional question are important to the application of international law to the settlement of any type of dispute: direct assertions of jurisdiction to apply policy without any intermediaries, and secondary assertions of competence in the form of recognition and enforcement of prior judgments and other determinations. Any consideration of environmental dispute settlement must recognize the necessity of the latter aspect of continued effective control as well as the former process of initial application. Consequently, this chapter examines the various procedural alternatives in the light of both jurisdicional dimensions. The more informal or 'political' processes of negotiation, good offices, mediation, and inquiry-and-conciliation, as well as the more formal or strictly 'legal' procedures of arbitration and judicial settlement, have pronounced utility – the choice of means varying according to the nature of the relevant parties and interests. In the case of reliance on judicial means, both international tribunals and national courts and administrative proceedings offer significant possibilities for peaceful and meaningful settlement of international environmental disputes. Finally, a question permeating much of the debate on international dispute settlement is that of the 'compulsoriness' of prior commitments to adjust grievances in certain specified ways, and this chapter therefore concludes with some observations and recommendations on that and some related matters in the context of environmental policy-making.

This may sound very familiar to those already well acquainted with the characteristics of international legal action, and indeed there is little that is unique about the processes and procedures for the settlement of international environmental disputes as distinct from other types of transnational controversies. There are nevertheless two important reasons for the inclusion of this discussion.

First of all, while highly experienced international lawyers might be tempted to skim through to the concluding pages, it seems necessary to give those newer either to the environmental field or to the practice the opportunity for a more complete understanding of what is actually going on in regard to an important aspect of the application of the law discussed previously and of certain features and difficulties endemic to the world constitutive process itself.

Secondly, dispute settlement in general and environmental dispute settlement in particular are currently very popular topics among both practitioners and scholars. Whereas the 1958 Law of the Sea Conference drafted only a short Optional Protocol on the settlement of international disputes, for example, delegates at UNCLOS III are now working out a whole elaborate system of dispute settlement. The present text contains an entire separate section of the proposed new 'umbrella treaty' consisting of nineteen separate draft articles on this subject. There are also four related annexes: on conciliation, on the Statute of the Law of the Sea Tribunal (a new body to be created under the treaty), on arbitration, and on special arbitration procedures for dealing with certain types of disputes (including those relating to protection and preservation of the marine environment).[1] This complex blueprint is, however, not yet in final form. Furthermore, as far as scholarship is concerned, the American Society of International Law has sponsored work on a series of essays for a volume on the avoidance and adjustment of environmental disputes,[2] which at the time of this writing has not yet appeared in published form. In view of this manifest interest, it therefore seems worthwhile to explore and expand upon organizational problems of international environmental dispute settlement.

Information availability

Accurate and broad-based information is critical to the rational management of environmental resources according to sound scientific principles. It is *ipso facto* crucial to the settlement of environmental disputes in a manner consonent with the comprehensive management regime.[3] From the point of view of the fulfilment or satisfaction of common environmental interests, there must be a central and relatively unbiased source of up-to-date and readily assimilable information on the present state of the environment and on variables likely to affect its continued protection and preservation. It follows that, however limited may be the ability of the United Nations Environment Programme or any other international organizations directly to carry out operational activities, the creation and smooth functioning of a centralized and co-ordinated information system – as envisioned by UNEP in Earthwatch[4] – must provide the foundation for a more ecologically sound world public order in the future. Beyond that, of course, when it comes to deciding between competing claimants or exclusive interests, exchange of or independent access to the relevant information is essential.

Types of international environmental disputes

Although there is an increasingly pressing environmental dimension to virtually every question in the global decision-making process today, obviously not every interaction necessitates or is even appropriate for an *ex post facto* adversary accounting of environmental costs and benefits. Situations which do, however, crystallize into environmental disputes may arise between governments, governments and international organizations, governments and private parties, or exclusively private parties. Because of the restrictions inherent in the nature of the world political system, the Statute of the International Court of Justice,[5] the Hague Conventions for the Pacific Settlement of Disputes,[6] the General Act for the Pacific Settlement of International Disputes,[7] the European and Pan American treaties on pacific settlement,[8] various multilateral and bilateral friendship treaties,[9] and other factors and agreements, the official parties to the vast majority of transnational dispute settlements are nation-states and occasionally international organizations. But the real parties in interest may well be corporations or other private business associations, public interest or citizens' groups, or private interests and individuals; and they are not infrequently the actual parties in the proceedings. The multiplicity and complexity of combinations that can result and the principle of economy dictate that settlement procedures and institutions should be invoked at the most immediate effective level of inclusivity. This means that where the causes and consequences of a controversy occur within its territorial boundaries and a particular state therefore has the authority and control over the pertinent resources, events, and persons, the appropriate competence is national (or local, according to the venue provisions of that state). Yet vast scope is thereby left for international concern and competence.

International environmental disputes can arise in connection with the use of either shared resources or non-shared resources; the consequences may also be diverse. It might be well to give some illustrations of environmental concerns which are of an international nature and types of disputes they can precipitate, but the following list is by no means either homogeneous or exhaustive. First of all, actions in or under the authority or control of one or some states may have transnational effects directly injuring the exclusive interests of other state(s). The circumstances in the *Trail Smelter* and *Gut Dam* arbitrations,[10] the *Nuclear Tests* cases,[11] the *Torrey Canyon* and Cherry Point oil spills,[12] and perhaps in the case of industrial wastes from Western Europe causing acidic rainfall in Sweden[13] provide a few random examples. Secondly, certain practices by states or their nationals or activities within their jurisdiction or control may have deleterious consequences for shared resources and therefore inclusive interests. The US project West Ford[14] might unfortunately have presented the textbook case, but all of the examples above could probably have been placed in this category as well. Thirdly, environmental programs can have potential for the diversion or distortion of international trade, capital flows, and development assistance. The US auto emission standards, for example, apply to imported cars as well;[15] the World Bank and the US Agency for International Development environmental criteria for lending and grants also affect capital flows.[16] And other examples are legion. Any or all of these and many other types of situation have resulted in or are capable of generating international environmental disputes.

Primary dispute settlement procedures

'Primary' or 'direct' jurisdiction is, to reiterate, the assertion of competence to apply policy directly to a person or thing, without the intrusion of any intermediary decision-makers. Such assertion is based on authority, a set of shared, conditioned subjectivities which provide the parties concerned with an indication of appropriate behaviour. Such assertion typically also rests on some degree of effective control, that is, a capability that can be employed to secure the desired result. But the latter is not always the case, and 'secondary' or 'indirect' assertions of competence may be required for the recognition and enforcement of authoritative decisions.[17] In assessing the available procedures for environmental dispute settlement, I intend to focus on primary applications of law.

Besides judicial settlement, Article 33 of the United Nations Charter mentions several other methods of pacific settlement of disputes.[18] The approach here will be to define and discuss the procedures of negotiation, good offices, mediation, inquiry-and-conciliation, arbitration, and judicial settlement as they lend themselves to resolution of different kinds of international environmental dispute. The last two are the means where decision is based most explicitly on legal rules, but international environmental law, understood as the expectations of the world community, is fundamental to the adjustment of differences through invocation of the more informal or political processes as well. All these distinctions are rather arbitrary, and in practice there is often a melding of several of the processes in any given situation or chain of events. Since the traditional classifications do, nevertheless, serve to highlight the varying functions that may need to be performed, they at least make useful summary headings here.

INVOCATION OF POLITICAL PROCESSES
Most environmentally relevant international disputes are resolved by political means. (In fact, while publicists keep referring to *Trail Smelter* as a milestone, their search for similar juridical decisions has revealed only the two of rather marginal significance, *Corfu Channel* and *Lac Lanoux*, to which *Gut Dam* ought to be added.)[19] This trend may be expected to continue, as the prevalent post-second world war practice with regard to the responsibility of states or the law of international claims has distinctly been settlement by lump-sum transfers between pairs of nations.[20] Flexible political approaches to the determination of international responsibility have much to recommend them in the light of the rapidly changing nature of environmental problems and perspectives thereon in the world decision process today. There are also, however, characteristic attendant disadvantages.

Negotiation
One of the speediest and most effective methods of resolving international disputes or potential disputes is negotiation.[21] Through deliberation, discus-

sion, conference, or other means of exchange of views, usually via normal diplomatic channels, the terms and conditions of a bargain are reached by mutual accord. Some form of negotiation must accompany any sustained association between persons or authorities not subordinated to one another, and environmental interactions are no exception to the rule.[22] In fact, given the enduring nature of most environmental problems and the continuous tort claims which they can generate, ongoing environmental negotiations between states are essential. International environmental law is a process of continuous demand and response, in which competing or conflicting environmental claims are put forth; and decision-makers must appraise and evaluate and accept or reject them in whole or in part. Negotiation has been defined as the first formal step.

Given geographic, geologic, and ecologic patterns and the resultant localized nature of most environmental disputes, bilateral negotiations are a likely approach. The dispute between the United States and Mexico over the salinity of the Lower Colorado River and the long drawn-out exchanges in the dispute provide an example of both inherent strengths and inherent weaknesses in this mode of settlement. After nearly half a century of controversy and several bitter recriminations, these two countries finally managed in 1973 to conclude their Agreement on the Permanent and Definitive Solution to the International Problem of the Salinity of the Colorado River.[23] Similar environmental disagreements and ongoing negotiations are characteristic of relations between the United States and Canada,[24] Poland and Czechoslovakia,[25] and many other pairs of countries regarding pollution and other problems of boundary waters and comparable types of environmental controversies. Some dispute situations, however, must be dealt with multilaterally. River and fisheries commissions are noted for providing arenas for negotiation on a multilateral scale for these purposes.[26] And the OECD principles and confrontation procedures for reporting national actions with international environmental impacts and dealing with transfrontier pollution are designed to encourage wider exchanges.[27] For a final example, almost the entire United Nations membership is participating in the current Law of the Sea Conference, and while the primary function of this assemblage is supposed to be the codification and progressive development of the law, its subsidiary role as a forum for the negotiation of already extant points of dispute must not be neglected.[28]

Although it has certain obvious merits in terms of maximum flexibility, the negotiation procedure has some inherent disadvantages for dispute settlement in general and environmental dispute settlement in particular. It fails to activate any machinery for the objective assessment of the situation, and there is no opportunity for states who are not parties to the immediate controversy to express their views. Furthermore, the relative bargaining strengths of the states involved based on factors extrinsic to the incident dispute tend to influence the proceedings. Because of the complex nature of environmental issues in the contemporary world decision process, these are serious liabilities for a means of dispute settlement from the perspective of the overall planning and management of the total human environment. In sum, the dispute may well be localized while the problems are much more internationalized.

Good offices, mediation and conciliation
The procedure of good offices consists in the attempt by one or more governments or eminent citizens who are not parties to the controversy to bring the disputants or potential disputants together, so as to make it possible for them to reach an adequate solution between themselves.[29] This procedure is an appendage of negotiation, its only real distinguishing trait being the introduction of a third party into the initial phase of the settlement process; it does not inject an additional point of view into the final outcome directly, as the performance of good offices is characteristically terminated with the assembling function without the third party taking an active role in the negotiations that result from its intercession. Some publicists regard mediation as part of the same procedure as good offices, while others distinguish it as a separate process implying a more active effort by the third party, who continues as go-between during the process of resolution and may offer proposals and counsel compromises.[30] That adds an element of sustained pressure of outside public opinion nudging the disputants towards agreement, but since it does not provide for any direct representation of broader community interests in the final accord it is not of distinct interest in the present context. By comparison, the procedure of inquiry-and-conciliation – sometimes called investigation-and-conciliation and other times split into its two component parts – consists in the submission of the controversy to a commission to find facts and make a report containing recommendations for settlement, although in theory each adversary party remains free to accept or reject these recommendations as it chooses.[31]

Without exessive drawing of fine lines, all these political processes have pronounced utility at least as a first resort in the settlement of differences of opinion in environmental matters. Particularly in cases where environmental controversies develop as a result of activities whose primary focus is far removed from ecology, such as the long-term effects of herbicides and defoliants employed in military operations, the use of good offices obviously suggests itself as a helpful tool. For another example somewhat less extreme, where there are many diverse factors at stake and negotiations seem inescapably complex and prolonged, as in the determination of rational utilization of waters of an international river, a mediator may clearly have a highly significant role to play in facilitating and expediting agreement. Finally, in disagreements over the interpretation and implications of a large body of scientific and technical data – for example, in the dispute that arose between Argentina and Brazil over the latter's proposed hydroelectric utilization of the Parana River[32] – one might well recommend a conciliation commission for the provision of a reasonable assessment; resort to such commissions is already mandated in several fisheries agreements,[33] and at UNCLOS III states have agreed to settle by compulsory conciliation disputes over alleged abuse of coastal state rights over living resources in the exclusive economic zone.[34]

The need for these various informal or political procedures was acknowledged at Stockholm, and it was understood that the Executive Director of UNEP would as necessary perform or provide for their performance.[35] They are in fact vital, and their summary discussion here should only be taken to indicate that their advantages as means of dispute settlement are not particularly distinctive to the environmental context. The current text for the proposed 'umbrella treaty' before UNCLOS III, it might be mentioned in passing, contains an article providing that any state which is party to a dispute relating to the interpretation or application of the convention may invite the other(s) to submit the dispute to conciliation, and, to repeat, it will contain another provision for compulsory conciliation for disputes relating to certain coastal state sovereign rights in the economic zone; in addition, there is a highly detailed annex setting forth the procedures to be followed for conciliation proceedings.[36] What is not discussed in that section or yet elsewhere is just what kind of substantive disputes are to be treated in the manner so elaborately delineated.

Furthermore, it is, of course, a general principle that an obligation to arbitrate or adjudicate becomes operative only after the exhaustion of these anterior remedies (although in practice a fair amount of latitude has been demonstrated in the interpretation of this requirement and the parties retain the right to choose to resort to judicial means).[37] And yet, all these diplomatic processes share certain basic deficiencies – primarily their lack of solicitation for broader inclusive common interests. When situations precipitate environmental disputes of a transnational nature, the underlying values at stake are almost by necessity more internationalized than the formalized scope of the controversy. This is so at least in great part because what the participants designate as culminating environmental events requiring the invocation of international dispute settlement mechanisms would most frequently be perceived by an outside observer as manifestations or consequences of wider patterns of behaviour. Often, also, these informal processes prove unsuccessful.

APPLICATION OF RULES OF LAW

The settlement of disputes by the application of rules of law to the facts is supposed to be the distinguishing feature of arbitration and adjudication. Although this definition clearly embodies an excessively narrow conception of the nature of international law and politics, it is reasonable to assert that these two dispute settlement processes are qualitatively different in that they anticipate more conscious resort to formalized or pre-established principles and rules by an outside tribunal. In the absence of clear specification in a treaty or some other form of agreement, there may be difficult preliminary questions of what rules of law are to be applied (with doctrines of *lex loci contractus, lex loci delicti,* and the 'public policy' of the forum figuring in the choice),[38] but the underlying assumption remains that there is some common core of expectations that can be drawn upon to order the relationships of the parties. The actual substance of international environmental law, including its reflection in certain opinions of international tribunals, has already been discussed at length in previous chapters. Consequently, the main purpose of this chapter and particularly of this section is briefly to give those unfamiliar with international juridical processes some idea of the basic steps and procedures that may be involved and of their practical implications. Many people tend to think of law only as that which is applied by courts, which should by now have been demonstrated to be an unworkably narrow notion. Although there is already a considerable and rapidly growing body of international law pertinent to environmental problems, to repeat, there have been very few international decisions directly or even marginally on point.

Arbitration

The Hague Convention of 1907 defines the goals of international arbitration as 'the settlement of disputes between States by Judges of their own choice on the basis of respect for law.'[39] Arbitration is primarily a creature of contract, and for international arbitration the statute that sets out the stipulations of the parties (as to the subject of the dispute, the powers and limitations of the tribunal, the procedures to be followed, the mode of defraying expenses, and so on) is usually called the *compromis.*[40] The Hague Conferences of 1899 and 1907 and their resulting Conventions on the Pacific Settlement of Disputes[41] created a ready arena in the form of the Permanent Court of International Arbitration, wherein a list of arbitrators is empanelled to facilitate selection by disputant states. And the rules of the PCIA were amended in 1962[42] to provide for arbitration and conciliation between two parties only one of which is a state, the other ranging from international organizations to private interests. Many pairs of states have concluded bilateral arbitration treaties and set up arbitral commissions of various sorts, and provision for resort to arbitration has also been included in some multilateral agreements, including environmental agreements.[43]

In the environmental context, by far the most famous arbitration is, of course, *Trail Smelter*,[44] whose basic chain of events is as follows. In 1928, the matter was first referred 'for examination and report' by the United States to the International Joint Commission, a bilateral investigative and juridical agency created by the Boundary Waters Treaty of 1909. The IJC report recommended payment by Canada of $350,000 to cover claims for damages incurred up to the close of 1931.[45] Later, under the terms of a convention signed by the two governments on 15 April 1935,[46] it was agreed to establish an arbitral tribunal (to consist of a neutral chairman and two national members) to decide questions of subsequent indemnity and a future regime for the operation of the smelter. In its decision reported to the two governments on 16 April 1938,[47] the Arbitral Tribunal found further damage between 1932 and 1937 and fixed indemnity for this at $78,000. In its second and final decision reported on 11 March 1941,[48] the same tribunal concluded that there had been no additional damage and it also set forth the 'prescribed regime' intended to 'remove the causes of the present controversy and ... result in preventing any damage of a material nature occurring in the State of Washington in the future.'[49] As provided for in the *compromis*, the tribunal reached its decision '[u]nder the principles of international law, as well as the law of the United States.'[50] The awards were paid in full.

What happened in *Lac Lanoux*[51] was not very different procedurally, although the decision itself was in favour of the defendant and so there was no award of damages. When the border between France and Spain was settled by the Treaty of Bayonne of 26 May 1866, the treaty was accompanied by an Additional Act providing for enjoyment of waters of common use. The dispute now in question arose thereunder in connection, it will be remembered, with French proposals for hydroelectric utilization of the Carol River (which drains from Lake Lanoux). In accordance with a 1929 bilateral Arbitration Treaty, France and Spain signed a *compromis* on 19 November 1956.[52] The resulting five-member Arbitral Tribunal (three neutrals and one French and one Spanish national) met and on 16 November 1957 announced its judgment. Again, while concluding that the proposed project would not be in violation of the international legal responsibility of France, the Tribunal stressed the obligation of an upstream state to reconcile the interests of other riparians with its own[53] – and environmentalists call special attention to the latter aspect of the decision.

In *Gut Dam*,[54] the flooding and erosion damage, which US citizens believed to have been caused at least in part by Canada's dam, occurred in 1951–2. In 1962 the US Congress enacted Public Law 87–587[55] authorizing the Foreign Claims Settlement Commission of the United States to adjudicate claims of US citizens against the government of Canada for the damage in question. Prior to the completion of the Commission's Lake Ontario Claims Program, extensive diplomatic negotiations between the two countries resulted in an agreement to establish the Lake Ontario Claims Tribunal. The arbitral *compromis* was signed on 25 March 1965,[56] and on that date the US Foreign Claims Settlement Commission, in accordance with the terms of Public Law 87–587, ceased processing the claims filed wih it.[57] The Agreement entered into force on 11 October 1966, and the first meeting of the Tribunal (which again consisted of a neutral chairman and two national members) was held in January 1967. The Tribunal handed down its first decision in favour of the United States on 15 January 1968,[58] holding that possible Canadian liability ran to any US citizen. On 30 January 1968, the Tribunal reconvened to consider the question of whether there was a time limitation on Canadian liability, and it gave its second decision against Canada on that point two weeks later, on 12 February.[59] At that time, the arbitrators suggested to the government agents involved that an attempt be made, 'without prejudice either to the validity of the claims or to the issue of substantive liability,'[60] to reach a negotiated settlement. Thereafter negotiations were undertaken and a compromise agreed upon,[61] which resolution was ratified by the Tribunal in its final communication of 27 September 1968.[62] The principle amount of all US claims had been $653,386.02, and Canada paid $350,000 in full settlement.[63]

Arbitration, on the whole, is a most useful technique for resolving international as well as other disputes. Its outcome may be based not only on international law, but also on the relevant legal provisions of the place(s) of injury. Thus, it both affords a means for the implementation of world community policies as embodied in various agreements and customary international law and also strengthens the ability of states further to protect their own and the common environment through the exercise of national prescriptive competence. It is, in addition, a more flexible process than most international adjudication, as in effect the parties themselves usually settle preliminary jurisdictional and choice of law questions and leave to the arbitrators the determination of substantive rights and responsibilities. Recognizing these benefits, delegates at UNCLOS III, for example, have carefully delineated detailed arbitration procedures for the resolution of disputes under the new law of the sea treaty.[64]

There are, nevertheless, certain important limitations to this mode of proceeding. Arbitration lacks the adaptability and pragmatism of political settlements, and the parties often encounter difficulties and delay in reaching agreement on arbitrators, means of proceeding, and so forth. When one party refuses even to acknowledge the existence of debatable issues, as was the position of Brazil in the Parana River controversy,[65] it is unlikely to be a viable option at all. And, as should be apparant even from the few examples above, the overall process may be drawn out for a period of twenty years or more, with the eventual award if any going to the children or grandchildren of the injured parties. Overall, despite this, the process of applying international law through arbitral procedures offers considerable hope for dealing with environmental situations of a relatively non-critical nature in such a way as to promote the best interests of the common heritage as well as adjust the incident claims of specific disputants.

Adjudication

In considering the application of international law through adjudication, it is necessary to examine the competences both of international judicial tribunals and other instruments and of national courts and administrative agencies. As far as the former are concerned, alternatives include application to the International Court of Justice, invocation of special existing arrangements, and proposals for new specialized international tribunals. As to the latter, a growing number of transnational disputes are being adjudicated in national arenas, and there is increasing recognition that in such instances the national bodies are effectively functioning as international decision-makers. In both cases the tribunals are applying the sources of international law enumerated in Article 38 of the Statute of the International Court of Justice,[66] and in both cases the resulting authoritative decisions themselves become part of international law.

International tribunals

One naturally thinks first of the 'principal judicial organ' of the United Nations, the International Court of Justice. Concerning environmental problems and other challenges of this technological age, former ICJ Judge Philip Jessup has outlined some potentialities already existing in the Court's Statute of which states may take advantage for improved dispute settlement.[67] He pointed out, for example, that the Court might try sitting in small groups called chambers instead of *en banc* and that it can provide for assessors with special technical competence to sit with it or one of its chambers – both of which have readily apparent merits to recommend them for dealing with specialized, often highly technical environmental controversies.[68] Judge Jessup also noted that some environmentally oriented conventions have made provision for compulsory jurisdiction in the World Court, and he seemed to find them more logical than those which contain elaborate specifications for special types of commissions.[69] While the international community could undoubtedly make fuller use of the ICJ as an environmental tribunal, there is, as Judge Jessup acknowledged, a serious drawback to the proposals: only states may be official parties in cases before the Court.[70] Aside from that constitutional restriction, other difficulties may be found as either causes or results of the generally low repute enjoyed by the World Court today.

The cases with most direct environmental relevance that have thus far come before the ICJ have been somewhat disappointing, both considering the course of their adjudication and as evidenced by their immediate practical outcomes (or lack thereof). Aside from the attempt at injunctive relief which has already been commented upon,[71] from the procedural point of view their most notable characteristic has probably been the lack of appellants before the Court.

Returning to the *Corfu Channel* case,[72] it will be remembered that after two British destroyers had ben damaged by mine explosions in Albanian territorial waters, with much attendant loss of life and personal injuries, British minesweepers swept the channel against the clearly expressed wishes of the government of Albania. In view of the implications of the dispute for the maintenance of international peace and security, the UN Security Council recommended that the two governments involved refer it to the World Court.[73] The United Kingdom filed an application instituting proceedings against Albania about one month later.[74] Albania raised a Preliminary Objection to what it considered a unilateral application in the absence of acceptance by Albania of compulsory jurisdiction under Article 36 of the ICJ Statute, but the objection was rejected by the Court in its decision of 25 March 1948.[75] Immediately after delivering that judgment, the Court was notified by agents of both parties that they had reached a Special Agreement on the questions to be decided, and the Court the next day issued an order in which it placed on record that the Special Agreement now formed the basis of further proceedings.[76] After receiving pleadings from both sides and hearing argument on the case, the judges handed down their decision on the merits on 9 April 1949,[77] holding among other things that Albania was responsible for the mine explosions and the resultant damage and loss of life but reserving for further consideration the assessment of appropriate compensation. When it came time to deal with the compensation question, the Albanian government was absent and made no submissions, denying the jurisdiction of the Court with regard to the assessment of damages. The ICJ, after appointing a panel of experts to examine the United Kingdom's figures and claims, accepted the applicant's total amount of £843,947 and rendered judgment accordingly.[78] Albania refused to pay.

In the course of adjudication of the two recent pairs of cases, the track record is not much more impressive, given that Iceland in the *Fisheries Jurisdiction*[79] and France in the *Nuclear Tests*[80] cases both failed to enter an appearance. As with the *Corfu Channel* assessment of compensation judgment, the ICJ nevertheless proceeded to decide the cases, pursuant to Article 53 of its Statute. Article 53 provides that whenever one of the parties does not appear or fails to defend its case, the other party may call upon the Court to decide in favour of its claim. But before doing so, the Court must satisfy itself not only that it has jurisdiction, but also that the claim is well founded in fact and law. Since the *Fisheries Jurisdiction* and *Nuclear Tests* share with *Corfu Channel* the fact that they relate to vital military or economic interests in the eyes of the defendant states, this may be an important indicator of the future of the ICJ as an institution.

In the *Fisheries Jurisdiction* case, the United Kingdom instituted proceedings against Iceland by a letter to the Court of 14 April 1972, but by a letter dated 29 May 1972 the Court was informed that the government of Iceland was not willing to confer jurisdiction on the ICJ and would not appoint an agent to present its case; a similar exchange occurred with regard to an application by the Federal Republic of Germany a little over a month later. By judgments dated 2 February 1973,[81] the judges decided they had jurisdiction to entertain both cases, but on 14 January 1974 the Court decided not to have joinder of the two sets of proceedings (since, while the basic legal issues in each appeared to be identical, the positions of the two applicants differed, as did their formal submissions).[82] After having issued and continued Orders Concerning Interim Measures of Protection in both cases, which were repudiated by Iceland (although Iceland and Britain did manage to reach an interim diplomatic agreement),[83] the World Court handed down its two judgments aginst Iceland on the merits on 25 July 1974[84] – right in the middle of the Caracas session of the Law of the Sea Conference, which was groping towards consensus on two-hundred-mile exclusive economic zones.

The course of events in the *Nuclear Tests* cases is not procedurally very different (except, of course, for the outcome), although an interesting twist was added by the application of Fiji to intervene in both Australia's and New Zealand's cases against France. Both applicants separately filed applications on 9 May 1973, and on 16 May the French government made known that it considered that the ICJ was manifestly not competent in the cases, that it could not accept the Court's jurisdiction, and that accordingly it did not intend to appoint an agent in the proceedings (nor was such agent ever appointed subsequently). Also on 16 May Fiji's application was filed under Article 62 of the Statute of the ICJ. Had the Court not merely held unanimously that the application to intervene lapsed when the Court found itself not called upon to render a decision in the main cases,[85] there would have been a tangled procedural problem to be faced; since Fiji was not a party to either the General Act of 1928 or the optional clause of the Statute, it was therefore questionable whether there existed a sufficient jurisdictional link between that country and France to justify intervention.[86] Subsequently the injunctive orders were issued, but France defied them and went ahead with its atmospheric tests, as has already been related.[87] On 20 December 1974,[88] the Court decided not to decide. In a sense, the outcome was clearly in favour of France, in spite of – or maybe even in consequence of – the fact that no pleadings were filed by the French government, it was not represented at the oral proceedings, and no formal submissions were made by it to the Court.

The ICJ is, of course, not the only arena for international adjudication. There is a second type of tribunal with potential for resolving some types of international environmental disputes, as provided for in several international agreements and organizations such as GATT, the IBRD, and the EEC. In addition to setting forth a regime for international trade law, the General Agreement on Tariffs and Trade has techniques for dispute settlement between GATT members (this role of GATT is basically adjudicatory, but also reflects of shades of all the other procedural alternatives).[89] Although initial stress is on negotiation, consultation, and referral to panels of experts, a dispute can eventually be brought before the contracting parties as a whole for formal authoritative decision in accordance with the terms of the Agreement and having the force of law. This may well be the way to work towards resolving controversies that arise over the effects of domestic environmental protection measures on imports and other related disagreements concerning countervailing duties, non-tariff barriers, and so forth. Like GATT, the International Bank for Reconstruction and Development has had to participate in the settlement of disputes concerning the international flow of capital for economic development, which led to the preparation of the Convention on the Settlement of Investment Disputes between States and Nationals of Other States.[90] The Convention itself establishes a structure for arbitration rather than more formal adjudication, but it is similar to GATT in setting up a defined and continuing arena for applying rules of law to a particular type of international situation. Disputes or potential disputes in the area of development and environment might find resolution in this forum. As far as the advanced industrial states are concerned, on the other hand, through its development of 'confrontation procedures' and other arrangements, the Organization for Economic Co-operation and Development seems already to be taking upon itself an analogous function.[91] And there are other specialized international adjudicatory arenas, such as the Council of Europe's European Court of Human Rights,[92] which may prove to have some competence or provide some guidance for dealing with controversies over distribution of resources, population movement and control programs, rights to health and well-being, and so forth.

A third type of tribunal is under consideration by the Law of the Sea Conference. The Optional Protocol to the 1958 Geneva Conventions on the Law of the Sea stipulated that disputes arising under the Conventions are within the compulsory jurisdiction of the International Court of Justice.[93] The United States, among other governments, did not ratify this protocol but is now an outstanding proponent of the creation of a completely new court to deal with these matters. The US delegation and that of Malta early made proposals for a Law of the Sea Court or International Maritime Court with overall jurisdiction over disputes arising under the new treaty[94] – although disputes in respect of fishing and possibly other subjects might instead be referred to special commissions or alternative fact-finding bodies.[95] It is difficult to understand why US policy-makers, who have frequently shown considerable frustration and dissatisfaction with the ICJ, think that they would be any more comfortable with a new court constituted at this time and with its composition and procedures determined by present international decision-making processes. However that may be, the idea has been adoptd by UNCLOS III, which has drawn up a Statute of the Law of the Sea Tribunal,[96] defining its organization and competence. Among other things, it is to be composed of twenty-one 'independent members, elected from among persons enjoying the highest reputation for fairness and integrity and of recognized competence in matters relating to the law of the sea'; they are to be chosen in such a manner as to assure 'representation of the principal legal systems of the world and equitable geographical distribution.'[97] The jurisdiction of the new body shall comprise both all disputes and applications submitted in accordance with the new law of the sea convention itself and 'all matters specifically provided for in any other agreement which confers jurisdiction on the Tribunal.'[98]

In addition, of course, to explicit settlement arrangements described in the preceding paragraphs, existing UN Specialized Agencies and other organs, agencies, and programs already deal with specific aspects of environmental protection and preservation and could help resolve disputes in these areas. The United Nations Environment Programmme has the most general express mandate and naturally comes to mind first in this connection. Similarly, the Intergovernmental Maritime Consultative Organization, in particular its Marine Environment Protection Committee, might represent an appropriate forum for certain disputes regarding vessel-source pollution. The same could be said of the International Atomic Energy Agency in the field of nuclear pollution and its hazards to human health and well-being. There are many other such examples.

Lastly, drawing on experience from all these tribunals and their work, it is not inconceivable that states may at some time think of trying to standardize environmental dispute settlement procedures as a distinct class, either within the context of the United Nations Environment Programme or elsewhere. This does, however, seem unlikely, as there is an environmental dimension to virtually every decision in the world decision process today and consequently many diverse areas in which transnational environmental disputes may arise. In any event, there have been no strong proponents of creating an International Environmental Court.

National courts and administrative tribunals
International law, transnational law, or the law of nations is part of national law as well, and it must be ascertained and administered by the courts of justice and administrative agencies of all appropriate jurisdictions as often as questions of right dependent upon it are duly presented for determination. This is true both of treaties and of customary international law.[99] Furthermore, these national bodies have an especially significant role as appliers of the law since, if the domestic law of the appropriate jurisdiction provides adequate redress, a would-be international claimant must generally show exhaustion of such remedies before being entitled to government espousal of the claim in an international tribunal. This rule is founded upon principles of economy, localization of delict and remedy, and basic good faith, but inflexibly applied it has often proved a source of much dissatisfaction to claimants by making ineffectual national adjudication the exclusive available means of resolving transnational claims.[100] There are, on the other hand, very often compelling reasons for preferring or needing to seek settlement in national rather

than international arenas. The relevant pros and cons should be easier to understand through discussion of some actual cases and of practical problems being perceived and new procedural arrangements being adopted for national environmental litigation.

Think again in this regard of the Amchitka and Trans-Alaska Pipeline controversies described in the last chapter.[101] In each case, *Committee for Nuclear Responsibility* v *Schlesinger*[102] and *Wilderness Society* v *Morton*[103] respectively, environmentalist litigation was instituted in US courts to try to block the proposed action. In both cases the interests of Canada and Canadian nationals were directly involved and wider community policies were affected, and in the Amchitka case Japanese interests and wellbeing were also directly at stake. In view of the potent transnational implications, a Canadian citizen and a Canadian environmental organization sought to intervene in the pipeline controversy, and their request was approved by the court on grounds that 'the interests of the United States and Canadian environmental groups are sufficiently antagonistic in this litigation to require granting of the application for intervention.'[104] Neither of these cases can, of course, by any means be cited as a decisive environmentalist victory in its actual outcome, since the blast went off on schedule and the pipeline has been built. *Committee for Nuclear Responsibility* failed to achieve consideration on its merits in the US Supreme Court, but at least it got some kind of hearing; and in *Wilderness Society*, the conservation groups claiming the environmental impact statement was defective did succeed in delaying the project for a while and in initiating some further review.

Even with these rather environmentally inconclusive results, national courts fulfilled a vital function in upholding international environmental considerations in the above disputes as well as on many other occasions. First of all, the time element can be crucial. Although the pressures of the situation forced accelerated and probably inadequate adjudication in the Amchitka case, a US national court was the only conceivable forum in which environmentalist claims could have been raised quickly enough to allow them to be determined authoritatively and with enough control to make the judgment enforceable in time to prevent the explosion. Secondly, and perhaps equally important, national courts are the final authority in interpreting a country's own laws, and national legislation is an essential complement to international prescriptions for the comprehensive protection of the human environment. In the pipeline case, for example, although there were profound international consequences at issue, the legal claims raised arose under the National Environmental Policy Act and other US statutes.

In the *Michie* v *Great Lakes Steel* case,[105] noted in passing in the preceding chapter, thirty-seven residents of Canada filed their own action directly in US courts. Plaintiffs, residents of the La Salle area in Ontario, filed a complaint against three corporations which operate seven plants in the United States immediately across the Detroit River. They claim that pollutants emitted by the plants are noxious in character and their discharge in the ambient air violates various municipal and state ordinances and laws, that such discharges represent a nuisance, and the pollutants are carried by air currents onto their premises in Canada, causing damage to their persons and property. In addition, plaintiffs jointly seek $1 million from each defendant, presumably as exemplary or punitive damages, because the nuisances complained of are alleged to be 'willful and wanton.' As far as the theory of the case is concerned, one Canadian professor has explained that *Michie* was brought on the basis of common law nuisance rather than under the 1970 Michigan Environmental Protection Act 'because the act provides for remitting the parties to the vagaries of the appropriate administrative authority for a determination of the legality of the defendant's actions and provides for injunctive relief, but not for monetary damages.'[106] As to the manner of proceeding, he also outlined half a dozen practical reasons for not using the international solution as in *Trail Smelter* but resorting to private action – including the duration of the proceedings, inertia on the part of governments on both sides of the border, difficulties of co-operation between government and industry, the preoccupation of the neighbouring governments with broader problems of Great Lakes water quality, and the lack of enforcement powers of the International Joint Commission to implement its recommendations – all of which were seen to inhibit dispute settlement by the former process.[107] This is not to ignore the fact that private actions, too, have their grave difficulties – not only peculiarities of the local forum, but also lack of involvement of governmental authorities, out-of-court settlements which leave the basic problem unresolved, and so on; but in this case, at least initially, plaintiffs see their best advantage in the national route.[108] A US Court of Appeals has upheld an earlier District Court decision denying a motion by defendants to dismiss the case, and the litigation is not finished.[109]

Although this is not demonstrated by the above case, there may be formidable procedural barriers in the form of standing requirements and other preconditions to be overcome before gaining access to national courts, particularly those of a foreign nation. Besides, private parties seek to further their own private interests, and certain broader common interests may go unrepresented. Consequently, there has been some international attempt to harmonize and liberalize procedural regulations, of which the Nordic Convention on the Protection of the Environment[110] is the outstanding example. In that Convention, the nuisance which environmentally harmful activities entail or may entail in another contracting state is specifically equated with a nuisance in the state where the activities are carried out.[111] The member countries – Denmark, Finland, Norway, and Sweden – have agreed that each other's citizens shall have the right to bring before the appropriate court or administrative authority of the state in which the allegedly harmful activities are being carried out the question of the permissibility of such activities (including the question of measures to prevent damage) and claims for compensation for resultant injury. This is to be allowed, both initially and on appeal, 'to the same extent and on the same terms as a legal entity of the State in which the activities are being carried out.'[112] They have also provided for the appointment in each state of a special supervisory authority 'to be entrusted with the task of safeguarding general environmental interests insofar as regards nuisances arising out of environmentally harmful activities in another Contracting State.'[113] Such supervisory authority is to be granted the right to pursue legal or administrative proceedings in another state party if an authority or other representative of the general public interest in that other state could itself do so.[114] And other supplementary dispute settlement provisions are made as well, including one for interstate consultations and additional arrangements for submission to a conciliation commission.[115]

So, in overview, national tribunals are not only highly capable of settling certain types of environmental controversy but are virtually irreplaceable. Like their international counterparts, they have certain characteristic deficiencies from the point of view of primary competence to apply international environmental law, but some of these can be overcome. When it comes to dealing with international environmental disputes, no one settlement procedure or type of arena is universally or even usually preferable; the positive and negative factors must be balanced and the means tailored to each specific case.

Secondary jurisdiction for recognition and enforcement

With all of the above primary procedures, some authoritative decision-maker was asserting a competence to apply policy directly to persons or events. Except for some international courts and tribunals, such exercise of authority is usually accompanied by a degree of effective control over the parties and circumstances. Frequently, however, a competence to apply is in fact asserted in situations where there is ineffective or insufficient control on the part of the primary applier. In such cases, after the initial dispute resolution, some other entity is called upon to recognize and enforce the judgment – that is, to assert 'secondary' or 'indirect' jurisdiction.[116] This is usually accomplished by national courts, although at times it may be effectuated by legislative or executive acts of national governments or occasionally international organizations.[117] In any event, one or both of two basic processes take place: the vindication of the competence of the primary authority in its application (recognition) and ancillary assistance by means of various sanctioning techniques (enforcement). (Sometimes, of course, only the former is required, as there will then be voluntary compliance or else the decision will be self-executing.)

Very little is new about international environmental law and organization in this regard, except that problems already encountered in other fields are exacerbated. In the political arena, however inappropriate a rigid adherence to the act of state doctrine (according to which sovereign immunity is invoked for governmental acts)[118] may be in cases of expropriation and other traditional applications, it is even more so when environmental issues are involved. One of the primary deficiencies of environmental dispute settlement by political means identified earlier is the tendency to minimize the representation of wider or more inclusive community policies and interests. If any secondary arena asked to effectuate such resolutions must decline to make determinations on the merits giving effect to principles of international law, a valuable opportunity might be overlooked for promoting protection of the common environment while at the same time upholding the latitude of states to resolve their differences through mutually advantageous political accommodations. As to the processes of application of rules of law, World Court judgments are allegedly immune from lateral attack, although there are several examples such as *Corfu Channel* of derogation of the duty of compliance;[119] arbitral awards may be vitiated by claims of nullity at international law, yet the relatively few that have been repudiated by states have been those affecting strongest-felt national interests (primarily territorial boundaries);[120] and finally, actions of international organizations may in theory be challenged on the basis of *excès de pouvoir* or of *ultra vires* their charter, but such instances, while not non-existent, are not plentiful either.[121] The strongest opposition is encountered when 'vital interests' are involved, and this effect has already been observed in several cases in the preceding pages: military interests in *Corfu Channel* and *Nuclear Tests* and interests highly important to the national economic livelihood in *Fisheries Jurisdiction*.

As to when in actual practice there is likely to be repudiation of authority and non-compliance with authoritative decisions, an outstanding authority on this subject has summarized the political determinants:

Law is a delicate balance of authority and control. Although authority per se constitutes a factor toward compliance with authoritative decisions, there is no absolute or necessary correlation between authoritative decision and controlling application ... When an authoritative process of decision overreaches its own ambit of control, a subsequent controlling process of decision is asserted and effects a decision more consonant with the actual, though shifting, power balance prevalent in the community.

Repudiation or attempted repudiation of an authoritative decision does not invariably follow every decision that contravenes the interests of a particular controlling group. Adverse decisions that do not mean serious shifts in power allocation are usually accepted; their sting is not so intense as to blind the victim's recognition of the common benefits in a working decision process. Authority has ventured somewhat outside the perimeter of control but has not crossed that Rubicon beyond which its own effectiveness is imperiled.

Repudiation is almost automatic, however, when 'vital interests' are affected or the power balance is threatened ... The vital quality of an interest is of course subjective; it can be a matter of dispute within the group asserting it. The imminent gain or loss of any value can be deemed vital.[122]

As a most fundamental conclusion, what all this supports is the suggestion at the beginning of this chapter that primary emphasis must be on working out mutual agreements and other accommodations so as to prevent environmental disputes arising in the course of international or transnational interactions. Although it is essential that there be adequate and readily available procedures for the *ex post facto* adjustment of grievances, the main thrust of international effort should be directed at the resolution of potential conflicts before they crystallize into adversary confrontations. When parties do nevertheless become arrayed against each other, Article 33 of the UN Charter and community policy generally evinces a preference for initial appeal to political means. Where those prove ineffectual in the opinion of the parties themselves or where their outcomes violate wider community interests, it is possible to resort to arbitration or adjudication. The eventual utility of these processes may depend both on the nature of the values in question (how closely the state responsible sees the consequences as impinging upon its central and vital interests) and on the extent and effectiveness of the pressure that can be brought to bear to enforce compliance (that is, actual political, economic, military, or other power exercisable against the perpetrator under the circumstances). Consequently, while in theory it is all very well to work out elaborate, compulsory, and supposedly binding commitments to settle disputes in accordance with prearranged patterns and procedural requirements, in practice there is likely to be a tentative and groping reliance on 'the science of muddling through' – which may be organized, non-erratic, and in some measure predictable, but is 'muddling through' nevertheless.[123]

'Compulsory' provisions

Many of the above observations cast aspersions on the intense concern of certain scholars and diplomats with provisions for 'compulsory' resort to certain procedures.[124] Delegates at UNCLOS III have nevertheless carefully worked out for their draft articles provisions for such acceptance of the jurisdiction of an arbitral tribunal, the Law of the Sea Tribunal, the International Court of Justice, or any two or three of them; they have also provided for compulsory conciliation in cases of abuse of certain coastal state rights concerning living resources in the exclusive economic zone.[125] In theory, it is extremely difficult to be against compulsory international dispute settlement – or, more precisely, compulsory pacific settlement of disputes. Even when international law is incomplete or non-existent on the issues in question, a decision may be rendered *ex aequo et bono*.[126] The pervasiveness of environmental considerations and the inherent nature of world public order today, however, counsel against placing too much reliance on these supposedly mandatory and often also purportedly binding provisions and on expending excessive energies and bargaining strengths in promoting their formal acceptance.

Two types of reservation in national declarations accepting the compulsory jurisdiction of the International Court of Justice under Article 36 of its Statute may prove severely limiting in this as in other contexts: the self-judging 'domestic jurisdiction' and 'national defence' reservations. The much-decried US 'Connally Amendment' is an example of the former,[127] and the French reservation for 'disputes concerning activities connected with national defence' unsuccessfully raised in the *Nuclear Tests* cases comes in the latter category.[128] In addition, Canada has now provided a precedent of an explicitly environmental reservation. Concurrently with the passage of the Arctic Waters Pollution Prevention Act and the Bill to Amend the Territorial Sea and Fishing Zones Act, the Canadian government modified its Article 36 declaration to terminate its acceptance of compulsory ICJ jurisdiction with regard to disputes arising out of or concerning jurisdiction or rights claimed or exercised by Canada in respect of the conservation, management or exploitation of the living resources of the sea, or in respect of the prevention or control of pollution or contamination of the marine environment in marine areas adjacent to the coast of Canada.[129]

The effect of any of these reservations is magnified by the condition of reciprocity which has been read into Article 36 (which enables the disputant state which has declared the wider acceptance of compulsory jurisdiction to rely upon the reservations to acceptance laid down by the other party).[130]

None of this goes to say that compulsory dispute settlement provisions – whether they refer to adjudication in the ICJ or elsewhere, mandatory arbitration, or some other manner of proceeding – are valueless or totally inconsequential. Far from it. The very fact of voluntary prior agreement to seek resolution in this way at the very least raises the costs of subsequent non-compliance in terms of national prestige and social pressure. The acceptance of compulsory jurisdiction, an obligation to arbitrate, or some alternative anticipatory commitment also indicates a basic commitment to the substance of international law and the integrity of international organization. But still, as has been said, when vital national interests are affected, the costs of compliance may seem too great, and a state may insist on reserving a matter or a whole class of matters solely to its own discretion. Many such claims will be special in nature and will produce results damaging to common interests; France's continuation of nuclear testing in the atmosphere has been repeatedly condemned by the international community. In certain instances, however, a state may repudiate review by outside authorities because its own position is in advance of, rather than contrary to, the formalized development of international law. If the Canadian example represents such a vanguard reservation, it should be able to be withdrawn following upon successful conclusion of a new Law of the Sea Treaty.

So, at the very end, we are back once again to the original question of the identification of common interests and provision for their satisfaction, for which reliance solely upon the good will of nations and their voluntary self-denial has been found wanting. On disputed issues, states are more apt to conform their behaviour to community norms when there has been an authoritative decision than when there has not, because such a judgment itself carries a degree of compelling weight. But, in the final analysis, progress towards a more ecologically sound world public order remains critically dependent upon the reorientation of fundamental underlying community expectations – that is, on the progressive development of international law.

Notes

CHAPTER ONE

1 B. Ward and R. Dubos, *Only One Earth* 30 (1972). See also J. McHale, *The Ecological Context* (1970).

2 H. Sprout and M. Sprout, *Toward a Politics of the Planet Earth* 14 (1971). On the general subject of political ecology, see H. Sprout and M. Sprout, *The Ecological Perspective on Human Affairs* (1965) and Russett, The Ecology of Future International Politics, in *International Politics and Foreign Policy* 93 (J. Rosenau ed 1969). See also S. Brubaker, *To Live on Earth* (1972); Clemens, Ecology and International Relations, 28 *Int'l J.* 1 (1972); *Managing the Planet* (P. Albertson and M. Barnett eds 1972); *Resources and Man* (Nat'l Acad. Sci.–Nat'l Research Coun. 1969); *World Eco-Crisis* (D. Kay and E. Skolnikoff eds 1972).

3 For a discussion of all three convocations, see Knelman, What Happened at Stockholm, 28 *Int'l J.* 28 (1972).

4 For discussion and contrast of the extremes, see Miller, Doomsday Politics: Prospects for International Co-operation, ibid, at 121.

5 D. Meadows, D. Meadows, J. Randers, and W. Behrens, *The Limits to Growth* (Report to the Club of Rome 1972); cf M. Mesarovic and E. Pestel, *Mankind at the Turning Point* (2d Report to the Club of Rome 1974).

6 See generally R. Falk, *This Endangered Planet* (1971). See also B. Commoner, *The Closing Circle* (1971); P. Ehrlich and A. Ehrlich, *Population, Resources, Environment* (2d ed 1972). But see, eg, J. Maddox, *The Doomsday Syndrome* (1972); Beckerman, Economic Development and the Environment: A False Dilemma, *Int'l Conciliation* no. 586, at 57 (January 1972). See also G. Taylor, *The Doomsday Book* (1970).

CHAPTER TWO

1 For some discussion of the political dimensions, see, eg, Wolfers, The Actors in
   International Politics, in *Theoretical Aspects of International Relations* 83
   (W. Fox ed 1959); Herz, The Territorial State Revisited: Reflections on the
   Future of the Nation-State, in *International Politics and Foreign Policy* 76
   (J. Rosenau ed 1969); Riggs, The Nation-State and Other Actors, in ibid, at 90.
   On international legal personality, see, eg, P. Jessup, *Transnational Law*
   (1956); McDougal, The Impact of International Law upon National Law: A
   Policy-Oriented Perspective, 4 *S. Dak. L. Rev.* 25 (1959); 1 D. O'Connell,
   International Law 89 and passim (2d ed 1970); 1 M. Whiteman, *Digest of
   International Law* 1 and passim (1963).
2 *Report of the United Nations Conference on the Human Environment*, UN
   Doc. A/Conf.48/14, at 2,7, Principle 21 (1972) (hereinafter *Report*)
3 G.A. Res. 2997, 27 U.N. GAOR Supp. 30, at 43, U.N. Doc. A/8730 (1972). As regards
   the last-mentioned body, however, it has recently been decided that the Environ-
   ment Co-ordination Board should be merged with the U.N. Administrative Committee on Co-ordination (Restructuring of the Economic and Social Sectors of the
   United Nations System, G.A. Res. 32/197 Annex (9 January 1978), in 17 *Int'l
   Legal Materials* 235, 246 (1978)).
4 See The United Nations System and the Human Environment: Consolidated
   Document Submitted by the Administrative Committee on Co-ordination, U.N.
   Doc. A/Conf.48/12 (1972). UNEP has now made arrangements to try to keep track
   of and appraise these various activities.
5 Many non-governmental organizations were represented at the UN Conference on
   the Human Environment, and they even issued their own 'NGO Declaration' on
   environmental policies (*Environment Stockholm* 17 (U.N. CESI 1972)). Eleven of
   those NGOs also promulgated a separate 'Statement of Youth and Student
   NGOs' (ibid, at 19). One author has counted 237 transnational associations of
   various types that participated in preparatory activities for the conference
   (Feraru, Transnational Political Interests and the Global Environment, 28 *Int'l
   Org.* 31 (1974)). See also T. Wilson, *International Environmental Action*
   (1971), which discusses the activities both of many inter- and non-governmental organizations in environment-related fields.
6 For an impressive and dramatic illustration of the interplay of economic power
   and ecological danger in just one important area, see N. Mostert, *Supership*
   (1974), the saga of the supertanker.
7 Pope Paul's 1968 encyclical which condemns the use of contraceptives. Cf *Family
   Planning in an Exploding Population* (Father J. O'Brien ed 1968).

8 The literature on public goods and public choice is vast. The theory was originally
   set forth by Paul Samuelson; see, eg, his Aspects of Public Expenditure Theories,
   40 *Rev. Econ. & Stat.* 332 (1958). It was further developed by John Head,
   Public Goods and Public Policy, 17 *Pub. Finance* 197 (1962), among others.
   And it has been applied to the theory of groups by Mancur Olson, *The Logic of
   Collective Action* (rev ed 1971).
9 On the distinction between common and special interests see M. McDougal,
   H. Lasswell, and I. Vlasic, *Law and Public Order in Space* 148–9 (1963).
10 Ibid, at 150–1
11 Olson, supra note 8. He explained: 'Because of the fixed and thus limited amount
   of the benefit that can be derived from the "collective good" ... in the market
   situation, which leads the members of a market group to attempt to reduce the
   size of their group, this sort of collective good will here be called an "exclusive
   collective good" Because the supply of collective goods in nonmarket situations
   automatically expands when the group expands, this sort of public good should be
   called an "inclusive collective good"' (ibid, at 38). He also pointed out that
   whether a group behaves exclusively or inclusively consequently depends upon
   the nature of the objective the group seeks, not on any characteristic of the membership (ibid, at 39).
12 M. Olson, Beyond the Measuring Rod of Money, chapter 3 at 60–3 (unpublished
   ms in hands of author 1978)
13 The basic requirement of a public good, in other words, will be met: if any actor
   X_i in a group $X_1, ..., X_i, ..., X_n$ consumes it, it cannot feasibly be withheld
   from the other actors in the group – where the group may be as large as the
   whole world or as small as two actors. This is the traditional definition of a
   public or collective good as originally promulgated by Samuelson. Head, supra
   note 8, identified such a good as having two main characteristics: infeasibility
   of exclusion and jointness of supply. In *The Logic of Collective Action*, Olson
   suggested that his notion of an 'exclusive collective good' recognizes the fact
   that, at least within some given group, the latter criterion may not be a relevant consideration; while there can never be group indivisibility without a public good or externality, in other words, the latter can exist without the former
   (Olson, supra note 8, at 38 n58; see also Olson, supra note 12, chapter 3 at 64).
   The 'if' in the above definition is highly significant. *Once the good is produced*
   any restriction on its use by some person, say by charging a price, would be
   'inefficient' in the sense of reducing the total utility of members of the community. But there is still the problem of inducing the creation or preservation of
   public goods, which may require charging for their use or excluding non-payers
   (if possible and not too costly) to provide incentive. This factor will in turn

restrict their use to some less-than-optimum level. See A. Alchian and
W. Allen, *University Economics* 147–8, 245–7 (3d ed 1972). See also 16
infra.

14 On this subject viewed as an economic problem, see generally *Managing the*
Environment: International Economic Cooperation for Pollution Control
(A. Kneese, S. Rolfe, and J. Harned eds 1971). See also A. Freeman, R. Have-
man, and A. Kneese, *The Economics of Environmental Policy* (1973).

15 For discussion of exclusive patterns of wealth distribution in oceans fisheries, see
Christy, Fisheries: Common Property, Open Access, and the Common Heritage
(Resources for the Future, reprint no. 101, 1972). See also F. Christy and
A. Scott, *The Common Wealth in Ocean Fisheries* (1965); Christy, Alternative
Arrangements for Marine Fisheries: An Overview (Resources for the Future,
Paper no. 1, 1973); Miles, Organizational Arrangements to Facilitate Global
Management of Fisheries (RFF, Paper no. 4, 1974).

16 But see Loehr, Collective Goods and International Cooperation: Comments, 27
Int'l Org. 421 (1973).

17 Olson, supra note 8, at 1–2

18 Ibid, at 23–36

19 Ibid, at 48–50

20 Ibid, at 27–9, 33–5. What is optimum is defined in the sense of Pareto optimality.
Consequently, the optimal amount of a collective good for a group as a whole to
obtain, if it should obtain any, would be reached when the gain to the group was
increasing at the same rate as the cost to the group. See ibid, at 27–31.

21 Ibid, at 35

22 Russett and Sullivan, Collective Goods and International Organization, 25 *Int'l*
Org. 845 (1971). For Olson's comments on this article, see his Increasing the
Incentives for International Cooperation, ibid, at 866. Compare Ruggie, Collective Goods and Future International Collaboration, 66 *Am. Pol. Sci. Rev.* 874
(1972); Loehr, supra note 16.

   The underlying assumption of this approach, that states tend to behave
analogously to private individuals in a market economy, has been questioned. It
has been argued that features endemic to bureaucratic politics introduce distortions in behaviour. As the trends described in chapters 3 and 4 demonstrate,
at least as regards the type of issues being dealt with in this book such effects
may be assumed to be relatively insignificant for present purposes. In the making of international law and the design of international organization, states can
be seen to behave as would be expected from the rational actors model set forth
here. But on bureaucratic politics and the usefulness of the theory in understanding other aspects of international relations, see G. Allison, *Essence of*
Decision (1971); M. Halperin, *Bureaucratic Politics and Foreign Policy*

(1974); R. Hilsman, *To Move a Nation* (1967) and *The Politics of Policy Making in Defense and Foreign Affairs* (1971).

23 In fact, any organization or group will usually be divided into subgroups or factions. This fact does not weaken the assumption that organizations exist to serve the common interests of members, for the assumption does not imply that intragroup conflict is neglected. As Olson himself pointed out, the opposing groups within an organization ordinarily have some interest in common, or else why would they stay in the organization? The members of any subgroup or faction, however, often have a common interest of their own (which common purpose may well consist in defeating some other subgroup or faction) (Olson, supra note 8, at 8 n11). In this sense, international organization is no different from organization of any other kind.

24 An external economy or positive externality occurs when an action taken by an economic unit results in uncompensated benefits to others. External diseconomies or negative externalities mean that the action results in uncompensated costs to others. Both may be economies of production or consumption. On failures of the market system due to the public goods nature of many environmental services, see A. Freeman, R. Haveman, and A. Kneese, supra note 14, at 65–97. See also O. Herfindahl and A. Kneese, *Quality of the Environment: An Economic Approach to Some Problems in Using Land, Water, and Air* 4–9 (1965); A. Kneese, R. Ayres, and R. d'Arge, *Economics and the Environment: A Materials Balance Approach* 2–7 (1970). See generally Coase, The Problem of Social Cost, 3 *J.L. & Econ.* 1 (1960); Davis and Whinston, Externalities, Welfare, and the Theory of Games, 70 *J. Pol. Econ.* 241 (1962); Buchanan and Tullock, Externality, 29 *Econ.* 371 (1962).

   The terminological shift here from public goods to externalities is meant to facilitate clarity of presentation of political choices. In fundamental economic theory, the two are indistinguishable (since 'public' does not mean the good must be produced by some governmental body): 'In many casual discussions, public goods and externalities are distinguished. Flowers in the public park are in these discussions described as a public good (because the government provides them), whereas the same flowers grown in some private citizen's garden next to a well traveled street are called an "external economy" (because the benefits they provide to those who travel the street are a byproduct of efforts an individual may have taken with no interests in mind but his own) ... [T]here is no fundamental basis for distinguishing public goods from externalities, but only differences in contexts and institutional arrangements ...' (Olson, supra note 12, chapter 3 at 65).

25 See O. Herfindahl and A. Kneese, supra note 24, at 8–9. On the interplay of economic factors and legal considerations, see d'Arge and Kneese, The Econo-

mics of State Responsibility and Liability for Environmental Degradation, in *International Responsibility for Environmental Injury* (R. Stein ed forthcoming).

26 See 11–12 supra.

27 G. Calabresi, *The Costs of Accidents* (1970). Professor Calabresi distinguishes 'general deterrence,' which 'involves attempting ... to decide what the accident costs of activities are and letting the *market* determine the degree to which, and the ways in which, activities are desired given such costs,' from 'specific deterrence,' which 'involves deciding collectively the degree to which we want any given activity, who should participate in it, and how we want it done' (ibid, at 68–9). Again upon the rough observation that sovereign states tend to behave analogously to private individuals in a market economy in their dealings with each other (12–13 supra), these two approaches are characteristic of the world public order as well.

28 Calabresi, supra note 27, at 68–9

29 See A. Alchian and W. Allen, supra note 13, at 244–7.

30 Ibid, at 245–6. See generally Coase, supra note 24; for a comment on the Coase theorem, see Calabresi, Transaction Costs, Resource Allocation and Liability Rules, 11 *J. L. & Econ.* 67 (1968).

31 A. Alchian and W. Allen, supra note 13, at 147–8, 247. See also, G. Calabresi, supra note 27, at 26–31.

32 For elaboration of the public order concepts of minimum and optimum order, see M. McDougal, H. Lasswell, and I. Vlasic, supra note 9, at 157–65.

33 UN Declaration on the Human Environment, *Report*, supra note 2, at 2,4; Principle 1

34 R. Falk, *This Endangered Planet* (1971) 105–32. This is a landmark study, whose fundamental concerns are shared by the present author. At one point in the book, however, Professor Falk makes the flat assertion that '[a] world of sovereign states is unable to cope with endangered planet problems' (ibid at 105), and he proceeds to advocate his new united world order of ecological humanism. One is reminded, nevertheless, of the foreboding of Innis Claude: 'The state is, after all, *there* – overwhelmingly the most significant power unit on the international scene' (I. Claude, *Power and International Relations* 247 (1962)). Indeed, Falk himself, along with Cyril Black, has predicted that at the very least 'the sovereign state is here to stay for the rest of the century' (1 *The Future of the International Legal Order* vii (R. Falk and C. Black eds 1969)). It would therefore seem that the first question to ask must be if and how the present state system, with all its defects, can at least begin to cope with ecological unities and protect and preserve the human environment until or unless a new world order takes over.

CHAPTER THREE

1 'Resources' in all these disciplines share the same core meaning: potential values, capabilities or means of acquiring goods and services.

2 The Third United Nations Conference on the Law of the Sea (UNCLOS III or LOS III) has been in preparation since 1970, when the UN General Assembly decided to convene a conference which would deal with the establishment of an equitable international regime – including international machinery – for the area and resources of the seabed and with a broad range of related issues (G.A. Res. 2750 C, 25 U.N. GAOR Supp. 28, at 26, U.N. Doc. A/8028 (1970)). The preparatory work was undertaken by the UN Committee on the Peaceful Uses of the Sea-Bed and the Ocean Floor Beyond the Limits of National Jurisdiction (the 'Seabeds Committee'). A two-week organizational session of UNCLOS III met in New York from 3 to 15 December 1973, and its first substantive session was held in Caracas from 20 June to 29 August 1974. Another substantive session met in Geneva from 17 March to 9 May 1975. The fourth, fifth and sixth sessions were held in New York, respectively from 15 March to 7 May 1976, 2 August to 17 September 1976, and 23 May to 15 July 1977. The seventh session met in Geneva from 28 March to 19 May 1978 and it reconvened in New York from 21 August to 15 September 1978. At the present time, the latest version of the negotiating text is the Informal Composite Negotiating Text (ICNT), in 8 *Third United Nations Conference on the Law of the Sea: Official Records* (1977). Some changes have already been agreed to, however, since the compilation of the ICNT. Although there is a general mood of cautious optimism among delegates, it still remains uncertain when and even whether a new 'umbrella treaty' will emerge.

3 Declaration of the United Nations Conference on the Human Environment, in *Report of the United Nations Conference on the Human Environment*, U.N. Doc. A/Conf.48/14, at 2,7 (1972) (hereinafter *Report*)

4 *Sic utere tuo ut alienum non laedas* (use your own property in such manner as not to injure that of another) has been described as one of those 'general principles of law recognized by civilized nations' which the International Court of Justice is to apply by virtue of Article 38 of its Statute. 1 Oppenheim, *International Law* 346–7 (8th ed Lauterpacht 1955). The extent of the prohibition of the abuse of rights, however, is not at all certain, and it must be left to international tribunals to apply and develop this still controversial doctrine by reference to individual situations. See ibid, at 347.

5 For discussion of how equilibrium was achieved between the two parts of Principle 21 by governments at the UN Environment Conference, see Sohn, The Stockholm Declaration on the Human Environment, 14 *Harv. Int'l L.J.* 423, 486–93 (1973).

6 The UN General Assembly has passed a whole series of resolutions on this subject. A recent resolution on 'Permanent Sovereignty over Natural Resources' (G.A. Res. 3171, 28 U.N. GAOR Supp. 30, at 52, U.N. Doc. A/9030 (1974)) recalls no less than six previous General Assembly resolutions relating to its subject-matter.

7 The responsibility of states for injury to aliens and the broader doctrine of state responsibility for internationally wrongful acts has been the subject of a wealth of material. It can be approached through 8 M. Whiteman, *Digest of International Law* 697 *et seq*. (1967); cf W. Bishop, *International Law* 742–899 (1962). The International Law Commission has been discussing a set of draft articles on the latter. See Ago, (First) Report on State Responsibility, [1969] 2 *Y.B. Int'l L. Comm'n* 125, U.N. Doc. A/CN.4/217 and Add.1 (1969) and his subsequent reports, as well as the comments of the members of the commission. In a recent draft on state responsibility, it might be noted, the ILC defines as an international crime, *inter alia*, 'a serious breach of an international obligation of essential importance for the safeguarding and preservation of the human environment, such as those prohibiting massive pollution of the atmosphere or of the seas' (Report of the I.L.C. on the Work of its 29th Session, U.N. Doc. A/32/183, art. 19(3)(d) (30 August 1977)). More generally, see *International Responsibility for Environmental Injury* (R. Stein ed forthcoming). See generally chapter 6 infra.

8 'Property' is usually defined as the aggregate of rights to, interests in, or dominion over a thing. In general, there have been three classes of international property: *res privatae* (things the property of one or more individuals – in this case, of states), *res nullius* (the property of nobody), and *res communes* (things common to all). If not already the case, the environmental dimension with its all-pervasiveness has outmoded the second of these. That leaves the relevant division here between individual state property and some form of collective competence.

9 International law recognizes five principle modes whereby a state can acquire title to and exercise exclusive control over territory: discovery and occupation, conquest or annexation, accretion, prescription, and cession (H. Briggs, *The Law of Nations* 250 (2d ed 1952)). See M. McDougal, H. Lasswell, and I. Vlasic, *Law and Public Order in Space* 828–67 (1963). This principle of exclusive territorial sovereignty is fundamental to the *U.N. Charter*; see art. 2, para. 4.

10 'Internal waters' are defined as 'waters on the landward side of the baseline of the territorial sea' (Convention on the Territorial Sea and the Contiguous Zone, *done* 29 April 1958, [1964] 2 U.S.T. 1606, T.I.A.S. no. 5639, 516 U.N.T.S. 205, art. 5). These waters include 'historic bays' and bays with straight base or closing lines of less than twenty-four nautical miles, as well as ports and harbours (ibid, arts. 7–8). Roadsteads also come within the regimes of either internal waters

or the territorial sea (ibid, art. 9). These internal waters are legally assimilated to their adjacent land masses and are therefore deemed coastal states' 'own' resources. See M. McDougal and W. Burke, *The Public Order of the Oceans* 64 (1962). For the new emerging regime of internal waters, see ICNT art. 8, in 8 *Third UNCLOS: Official Records*, supra note 2, at 7.

11 The Convention on the Territorial Sea and the Contiguous Zone, supra note 10, art. 1, recognizes that '[t]he sovereignty of a State extends, beyond its land territory and its internal waters, to a belt of sea adjacent to its coast, described as the territorial sea.' There has been much dispute over the width of the territorial sea, but no country has traditionally disputed the contention that the marginal sea belonging to a coastal state embraces a belt of at least three nautical miles. The sovereignty that a state has been permitted to exercise over this area, however, has been subject to certain restrictions – most significantly, the right of 'innocent passage' (ibid, arts. 14–23). See also ICNT arts. 17–32, in 8 *Third UNCLOS: Official Records*, supra note 2, at 7–10.

12 Access to such airspace is entirely dependent upon the will of the subjacent state, which is free to decide unilaterally whether or not to admit foreign aircraft and under what conditions. See Convention on International Civil Aviation (the 'Chicago Convention'), *done* 7 December 1944, 61 Stat. 1180 (1947), T.I.A.S. no. 1591, 15 U.N.T.S. 295, art. 1; M. McDougal, H. Lasswell, and I. Vlasic, supra note 9, at 244–81.

13 The continental shelf within national jurisdiction is currently defined as the seabed and the subsoil adjacent to the coast but outside the area of the territorial sea 'to a depth of 200 metres or, beyond that limit, to where the depth of the superjacent waters admits of the exploitation of the natural resources of the said areas' (Convention on the Continental Shelf, *done* 29 April 1958, [1964] 1 U.S.T. 471, T.I.A.S. no. 5578, 499 U.N.T.S. 311, art. 1). The Continental Shelf Convention specifies that rights over the shelf are 'exclusive.' If a coastal state does not explore or exploit its resources, no one else may do so 'without the express consent of the coastal State' (ibid, art. 2). In the *North Sea Continental Shelf* case, [1969] I.C.J. 3, the International Court of Justice held that the basic provisions of this convention codified pre-existing customary international law. Thus, they reflect norms binding on all states rather than only the adherents to the treaty. For the new law of the continental shelf, see ICNT arts. 76–85, in 8 *Third UNCLOS: Official Records*, supra note 2, at 16–17.

14 The status of genetic, aesthetic, and cultural resources has traditionally been determined by that of the area or medium within which they are located. Those which are situated on land masses, therefore, have been treated as subject to the exclusive competence of the government possessing dominion over that territory. And the multitudes of flora and fauna that live in the sea have thus been deemed to be inclusive resources.

15 Ships of all states may traverse the high seas almost entirely free of prohibition by any state or group of states. See M. McDougal and W. Burke, supra note 10, at 767. The conventional regime regarding free use of the airspace above the high seas is very much the same. See International Convention Relating to the Regulation of Aerial Navigation, *done* 13 October 1919, corrected text in U.S. Dept. State Pub. no. 2143 (1944), and the 1944 Chicago Civil Aviation Convention, supra note 12. Since the seabed and the ocean floor have only very recently become available to direct access by man, there are few explicit prescriptions regarding competence over them. Their main non-military use has been for laying long distance communications cables, and no one has challenged their inclusivity for this purpose. See M. McDougal and W. Burke, supra note 10, at 780. The international regime for the deep seabed is, of course, a primary concern of UNCLOS III, and the Conference expects to change the whole pre-existing international legal order. See ICNT arts. 133–92 and Annexes II–III, in 8 *Third UNCLOS: Official Records*, supra note 2, at 22–34, 49–57.

16 The inclusive character of competence over the void of space and celestial bodies has been established by the past practice of states and by concomitant expressions of expectations by different community spokesmen. See M. McDougal, H. Lasswell, and I. Vlasic, supra note 9, at 226–7. A UN General Assembly resolution in 1961, for example, provided that outer space and celestial bodies 'are free for exploration and use by all States in conformity with international law and are not subject to national appropriation' (G.A. Res. 1721 A, 16 U.N. GAOR Supp. 17, at 6, U.N. Doc. A/5100 (1961)). See Declaration of Legal Principles Governing the Activities of States in the Exploration and Use of Outer Space, G.A. Res. 1962, 18 U.N. GAOR Supp. 15, at 15, U.N. Doc. A/5515 (1963). See generally the following outer space and disarmament resolutions: G.A. Res. 1348, 13 U.N. GAOR Supp. 18, at 5, U.N. Doc. A/4090 (1958); G.A. Res. 1472, 14 U.N. GAOR Supp. 16, at 5, U.N. Doc. A/4354 (1959); G.A. Res. 1884, 18 U.N. GAOR Supp. 15, at 13, U.N. Doc. A/5515 (1963). Equality of opportunity for all states in competence over and use of this resource was incorporated into treaty form in 1967 (Treaty on Principles Governing the Activities of States in the Exploration and Use of Outer Space, Including the Moon and Other Celestial Bodies (the 'Outer Space Treaty'), *done* 27 January 1967, [1967] 3 U.S.T. 2410, T.I.A.S. no. 6347, 610 U.N.T.S. 205).

17 The Statute on the Regime of Navigable Waterways of International Concern, *done* 20 April 1921, 7 L.N.T.S. 51, provides that each of the contracting states 'shall accord free exercise of navigation to the vessels flying the flag of any one of the other Contracting States' and provides further that '[n]o distinction shall be made between ... the different riparian States' or even 'between riparian and non-riparian States' (ibid, arts. 3–4). Where the shared use of international rivers conflicts with other activities of riparian states (eg, power, indus-

trial uses, dredging, clearing and maintenance), the principle of the 'community of interest of riparian States' governs. See *Case of the Territorial Jurisdiction of the International Commission of the River Oder*, [1929] P.C.I.J. ser. A., no. 23, at 27.

18 Due to divergences in historical development, there are differences in the traditional international legal status of the Arctic and Antarctic regions. The original pattern in relation to the Antarctic was and has remained inclusive, although that may now be changing (Antarctic Treaty, 1 December 1959, [1961] 1 U.S.T. 794, T.I.A.S. no. 4780, 402 U.N.T.S. 71). The Arctic land masses, however, have traditionally been regarded as sufficiently comparable to the other land masses of the earth to subject them to exclusive appropriation. Heretofore, despite certain Soviet claims to the contrary, national sovereignty over the Arctic basin has not generally been recognized. See M. McDougal, H. Lasswell, and I. Vlasic, supra note 9, at 792; see generally D. Pharand, *The Law of the Sea in the Arctic* (1973).

19 S. von Ciriacy-Wantrup, in *Resource Conservation: Economics and Policies* 35 (rev ed 1963), divides resources into two major groupings: renewable or 'flow' and non-renewable or 'stock' resources. McDougal, Lasswell, and Vlasic, supra note 9, at 779, add a third category of 'spacial extension' resources for 'those resources whose most distinct characteristic is their utility as media of transportation and communication.' They cite as examples land and ocean surfaces, airspace, and outer space (ibid).

'Flow' resources are themselves divisible into those which are not significantly affected by human action, eg, cosmic rays, solar radiations, gasses, winds, oceanic currents and tides, minerals dissolved in the waters, and gravitational and magnetic fields, and those which are so affected. The latter class is further subdivisible into those resources affected by human action which have a 'critical zone' (a range of occurrence and reproduction, which, if reduced below a certain level, will continue to decrease in an economically irreversible fashion), eg, pelagic seals, whales, and salmon, and those which do not, eg, waters of international rivers. See ibid, at 777–81. All these flow resources have traditionally been considered of shared ownership and open to inclusive regulation and use on the basis of equality of opportunity.

20 Minerals are the most important stock resources and have to date most often been mined within the territory of nation-states. The same conditions and decisions which have made the land masses and continental shelf subject to exclusive appropriation, therefore, also embrace these resources. On the right to exploit freely natural wealth and resources, see G.A. Res. 626, 7 U.N. GAOR Supp. 20, at 18, U.N. Doc. A/2361 (1952) and the 'Permanent Sovereignty' resolutions, supra note 6. But the picture is changing now that the minerals of the deep seabed are beginning to be exploited. See 59–60 infra and chapter 4, 86–7.

21 Supra note 10, art. 24

22 See M. McDougal and W. Burke, supra note 10, at 584–606.

23 This principle is interrelated with several other international law principles. It is similar to the 'nationality' principle, which confers competence on a state when its nationals are actors, and to the related 'passive personality' principle, which does so when its nationals are victims. It is somewhat narrower than the 'protective principle' underlying the various forms of self-help, since the latter extends a broad mantle of competence and use rights in relation to events affecting various vital interests of the state as a whole. See 46, 71–2 infra.

24 The historic function of the concept of contiguous zones to mitigate the legal consequences of the rigid regime of the territorial sea was summarized in a report done for the International Law Commission: 'The notion of the high seas zone contiguous to the marginal sea is a technical legal notion aimed at justifying special powers exercised by the littoral State on the high seas beyond the limit of the marginal sea' (Memorandum on the Regime of the High Seas 28, U.N. Doc. A/CN.4/32 (1950)). It has been noted that both 'contiguous' and 'zone' lose much of their reference to geographic factors and take on a reference to legal consequences – the consequences that certain exclusive claims are lawful in this context (M. McDougal and W. Burke, supra note 10, at 75 n164).

25 *Report of the I.L.C.*, 11 U.N. GAOR Supp. 9, at 4, U.N. Doc. A/5159 (1956), art. 3 of articles on the Law of the Sea

26 Ibid

27 Ibid

28 1 *Report of the Committee on the Peaceful Uses of the Sea-Bed and the Ocean Floor Beyond the Limits of National Jurisdiction*, 28 U.N. GAOR Supp. 21, at 49, U.N. Doc. A/9021 (1973)

29 For a synoptic table of laws concerning the breadth and juridical status of the territorial sea and adjacent zones in force in eighty-six states represented at the Geneva conferences, see *Second United Nations Conference on the Law of the Sea: Official Records* 157–63, U.N. Doc. A/Conf.19/8 (1960). A similar table of claims to sovereignty was prepared by some editors on the eve of UNCLOS III. See 1 *New Directions in the Law of the Sea* 833–69 (S. Lay, R. Churchill, and M. Nordquist eds 1973).

30 See ICNT art. 3, in 8 *Third UNCLOS: Official Records*, supra note 2, at 6. For confirmation, see Stevenson and Oxman, The Preparations for the Law of the Sea Conference, 68 *Am. J. Int'l L.* 1, 9 (1974).

31 See ICNT arts. 46–54, in 8 *Third UNCLOS: Official Records*, supra note 2, at 11–13. For an excellent discussion of the problem, see Oxman, Third United Nations Conference on the Law of the Sea: The 1977 New York Session, 72 *Am. J. Int'l L.* 57, 65–6 (1978). In considering the subject of islands, it is important to distinguish coastal or offshore archipelagos from outlying or oceanic archipelagos or archipelagic states; there are already some estab-

lished rules of international law applying to the former. See Evensen, Certain Legal Aspects Concerning the Delimitation of the Territorial Waters of Archipelagos, U.N. Doc. A/Conf.13/18 (1957); see also D. Pharand, supra note 18, at 69–88; compare Stevenson and Oxman, The Third United Nations Conference on the Law of the Sea: The 1970 Caracas Session, 79 *Am. J. Int'l L.* 1, 21–2, 24–5 (1975).

32 For discussion see Stevenson and Oxman, supra note 30, at 9–13; see also note 15 supra. The draft articles on 'transit passage' through straits used for international navigation are now found in the ICNT arts. 37–44, in 8 *Third UNCLOS: Official Records*, supra note 2, at 10–11.

33 See Text of the Declaration of Santo Domingo approved by the meeting of Ministers of the Specialized Conference of the Caribbean Countries on Problems of the Sea, held on 7 June 1972, in *Report of the Committee on the Peaceful Uses of the Sea-Bed and the Ocean Floor Beyond the Limits of National Jurisdiction*, 27 U.N. GAOR Supp. 21, at 70, U.N. Doc. A/8721 (1972); Conclusions in the General Report of the African States Regional Seminar on the Law of the Sea, held in Yaoundé, from 20 to 30 June 1972, ibid, at 73; Organization of African Unity, Declaration on Issues of the Law of the Sea (Addis Ababa Declaration), 2 *Seabeds Report*, supra note 28, at 4. By contrast, the United States, for example, did not announce acceptance of the concept until well into the Caracas session of UNCLOS III, and some other countries, like Russia and Japan, held out considerably longer.

34 For discussion of this point, see García-Amador, The Latin American Contribution to the Development of the Law of the Sea, 68 *Am. J. Int'l L.* 33 (1974). References are to Presidential Proclamation no. 2667, Concerning the Policy of the United States with Respect to the Natural Resources of the Subsoil and Sea Bed of the Continental Shelf, and Presidential Proclamation no. 2668, Concerning the Policy of the United States with Respect to Coastal Fisheries in Certain Areas of the High Seas, 59 Stat. 884 and 885 (1945), 13 *Dept. State Bull.* 485–6 (1945); and to the Declaration of Santiago, 18 August 1952, in *U.N. Legislative Series, Laws and Regulations on the Regime of the Territorial Sea* 723, U.N. Doc. ST/LEG/SER.B/6 (1957).

35 ICNT arts. 55, 58, in 8 *Third UNCLOS: Official Records*, supra note 2, at 13. For other elements of the overall 'package,' see Oxman, supra note 31, at 67–80. The rights of landlocked and other states with 'special geographical characteristics' are discussed infra 67, 70–1. As to the dispute settlement issue, the emerging solution now appears certain to be one of compulsory conciliation for abuse of certain specifically delineated coastal state rights. Many coastal states, however, still maintain that acceptance of the proposed compromise formulae on these last two subjects is conditional on agreement on the so-called Irish formula for determining the outer edge of the continental margin.

36 These general provisions and articles on global and regional co-operation, on technical assistance, and on monitoring and environmental assessment are ICNT arts. 193–207, in 8 *Third UNCLOS: Official Records*, supra note 2, at 35–6. See generally Schneider, Something Old, Something New: Some Thoughts on Grotius and the Marine Environment, 18 *Va. J. Int'l. J.* 147 (1977); Timagenis, Marine Pollution and the Third United Nations Conference on the Law of the Sea: The Emerging Regime of Marine Pollution (Center for Marine Law and Policy 1977).

37 The provisions on international rules and national legislation to prevent, reduce, and control pollution of the marine environment are ICNT arts. 208–12, in 8 *Third UNCLOS: Official Records*, supra note 2, at 36–7.

38 Art. 208(4), ibid, at 39

39 Art. 21(2) ibid, at 8

40 The environmental enforcement provisions of the ICNT are found in arts. 214–23, ibid, at 37–9.

41 Safeguards are currently found in arts. 224–34, ibid, at 40–1. See also the articles on responsibility and liability and on sovereign immunity, arts. 236–7, ibid, at 41.

42 Art. 235, ibid, at 40

43 R.S.C. 1970 (1st Supp.), c. 2 (1970). In this act, Canada declared the world's first 'anti-pollution' zone of one hundred nautical miles off its Arctic coast. Explicitly denying assertion of any new claims to 'sovereignty' in the area, at the time of its passage, then Secretary of State for External Affairs Mitchell Sharp explained the nature of the functional or zonal competence created by the act: 'The Arctic Waters bill represents a constructive and functional approach to environmental preservation. It asserts only the limited jurisdiction required to achieve a specific and vital purpose. It separates a limited pollution control jurisdiction from the total bundle of jurisdictions which together constitute sovereignty ...' (statement by the Honourable Mitchell Sharp, in the House of Commons on 16 April 1970, as quoted in 9 *Can. Y.B. Int'l L.* 284 (1971)). For an excellent comprehensive discussion of Arctic problems, see D. Pharand, supra note 18.

44 Supra note 13, art. 2. On the problems with other types of species, see Koers, Fishery Proposals in the United Nations Seabed Committee: An Evaluation, 5 *J. Maritime L. & Com.* 183 (1973).

45 On coastal species, highly migratory species, and sedentary species, see respectively ICNT arts. 61–3, 64 and Annex I, and 68, in 8 *Third UNCLOS: Official Records*, supra note 2, at 14–15, 15 and 49, and 15. See also art. 65 on marine mammals, ibid, at 15.

46 ICNT art. 66, ibid, at 15. It might be noted that there is a somewhat similar article on catadromous species (opposite of anadromous, ie, living in fresh water and going out to sea to spawn – eg, eels) (art. 67, ibid). For discussion and illustration

of the special problems of anadromous species, see R. Haig-Brown, *The Salmon* (1974).

47 *Report*, supra note 3, at 40. On weather and climate modification, see, eg, MacDonald, How to Wreck the Environment, in *Unless Peace Comes* (N. Calder ed 1968); Samuels, International Control of Weather Modification Activities: Peril or Policy? 13 *Natural Resources J.* 327 (1973); *Controlling the Weather* (H. Taubenfeld ed 1970).

48 Arts. 1, 5, G.A. Res. 31/72 Annex, 31 U.N. GAOR Supp. 39, at 36, U.N. Doc. A/31/39 (1976), reprinted in 16 *Int'l Legal Materials* 88 (1976). The convention originated in a proposal by the USSR containing a draft International Convention on the Prohibition of Action to Influence the Environment and Climate for Military and Other Purposes Incompatible with the Maintenance of International Security, Human Well-Being and Health (G.A. Res. 3264 Annex, 29 U.N. GAOR Supp. 31, at 27, U.N. Doc. A/9631 (1974)). Subsequently, the United States and the USSR got together on this problem and submitted identical drafts of a convention on the prohibition of military and other hostile use of environmental modification techniques (31 U.N. GAOR Supp. 27, Annex II (documents CCD/471 and CCD/472), U.N. Doc. A/10027 (1975)). See also G.A. Res. 3475, 30 U.N. GAOR Supp. 34, at 25, U.N. Doc. A/10034 (1975). For the bilateral accord, see Statement on Dangers of Military Use of Environmental Modification, July 3 and Text of Joint Communique, July 3, in 71 *Dept. State Bull.* 185 (1974). For discussion, see U.S. and Soviet Offer a Pact on Weather, *N.Y. Times*, 22 August 1975, at 3, col. 1.

49 Prohibition Convention, supra note 48, art. 2

50 G.A. Res. 626, supra note 20. But see Treaty Relating to Spitzbergen, *done* 9 February 1920, 43 Stat. 1892, T.S. no. 686, 2 L.N.T.S. 7. This arrangement concerning an 'exclusive resource' between the United States, the United Kingdom, Canada, Australia, New Zealand, the Union of South Africa, India, Denmark, France, Italy, Japan, Norway, the Netherlands, and Sweden is perhaps unique at international law. 'While recognizing the sovereignty of Norway over the Archipelago of Spitzbergen,' these countries agreed that 'ships and nationals of all the High Contracting Parties shall enjoy equally the rights of fishing and hunting in the territories [of the Spitzbergen archipelago] ... and in their territorial waters' (ibid, arts. 1, 2). The treaty now has forty parties. See also Report no. 39 to the Norwegian Storting concerning Svalbard (Royal Min. of Justice, abridged transl. 1974–5).

51 See note 6 supra. For discussion in the context of Principle 21, see Sohn, supra note 5, at 486–7 *et seq.*

52 See 21 supra.

53 *Done* 18 November 1974, in 14 *Int'l Legal Materials* 1 (1975)

54 See generally Review of the Impact of Energy Production and Use on the Environment and Role of the United Nations Environment Programme, U.N. Doc. UNEP/GC/31/Add.1 (1975); Bevan, Energy Development: A Need for Vision and Statesmanship, 12 *Alberta L. Rev.* 1 (1974); Dreyfus and Grundy, Influence of the Energy Crisis upon the Future of Environmental Policy, 3 *Env. Aff.* 252 (1974); Muir, Legal and Ecological Aspects of the International Energy Situation, 8 *Int'l Lawyer* 1 (1974).

55 For discussion of the efforts of US environmentalists to prevent SST landings, see *British Airways Bd.* v *Port Authority of New York and New Jersey*, 564 F.2d 1002 (2d Cir. 1977). See generally Larsen and Faggen, Regulation of Stratospheric Flights to Control Adverse Environmental Effects, 40 *J. Air L. & Com.* 259 (1974).

56 This set of prescriptions and practices is often known in the United States as the 'conflict of laws,' but in most other countries it is referred to as 'private international law.' It traditionally includes three general subjects: the bases on which a court will assert jurisdiction over the persons and the subject-matter, the effect to be given to judgments rendered by courts of other states, and the principles governing choice of law. See H. Steiner and D. Vagts, *Transnational Legal Problems* 79 (1968).

57 Provisions of 'responsibility of states,' also frequently referred to as those of an 'international minimum standard,' have proved difficult to define in a number of current efforts to 'codify' or 'restate' them. They constitute a sort of 'international bill of rights' as to the persons and property of aliens. See generally ibid, 275–425. Chapter 6 will discuss this at length.

58 *Done* 26 March 1975, [1975] 1 U.S.T. 540, T.I.A.S. no. 8056

59 *Done* 12 May 1954, [1961] 3 U.S.T. 2989, T.I.A.S. no. 4900, 327 U.N.T.S. 3; with amendments *adopted* 11 April 1962, [1966] 2 U.S.T. 1523, T.I.A.S. no. 6109, 600 U.N.T.S. 332; 21 October 1969, T.I.A.S. no. 8505; 15 October 1971, text in 11 *Int'l Legal Materials* 267 (1972)

60 *Done* 29 April 1958, [1962] 2 U.S.T. 2312, T.I.A.S. no. 5200, 450 U.N.T.S. 82. The High Seas Convention was one of four conventions adopted at the Geneva Conference on the Law of the Sea which met in March and April 1958; the other three, which were also done on the same day were: Convention on the Continental shelf, supra note 13; Convention on the Territorial Sea and the Contiguous Zone, supra note 10; Convention on Fishing and Conservation of the Living Resources of the High Seas, [1966] 1 U.S.T. 138, T.I.A.S. no. 5969, 559 U.N.T.S. 285. A second U.N. conference was convened in 1960 to settle the question left outstanding of the limits of territorial waters. The present assemblage is accordingly the Third United Nations Conference on the Law of the Sea.

61 *Done* 10 October 1957, in 37 *Dept. State Bull.* 759 (1957). The United States, among other governments, did not sign this convention. See 9 M. Whiteman, supra note 7, at 230–3. See also the new Convention on Limitation of Liability for Maritime Claims, in 16 *Int'l Legal Materials* 606 (1977); this convention uses a new unit of account, International Monetary Fund special drawing rights (SDRs).

62 25 May, 1962, in 57 *Am. J. Int'l L.* 268 (1963). This convention is not yet in force. For discussion, see 9 M. Whiteman, supra note 7, at 285–99. It is actually one of four treaties that have been negotiated to form the basis of liability for nuclear harm; the other three are: International Convention on Civil Liability for Nuclear Damage (the 'Vienna Convention'), *opened for signature* 21 May 1963, in 2 *Int'l Legal Materials* 727 (1963); Convention on Third Party Liability in the Field of Nuclear Energy (the 'Paris Convention'), *done* 29 July 1960, in 8 *Europ. Y.B.* 202 (1960); Convention Supplementary to the (OECD) Paris Convention of 1960, 31 January 1963, in 2 *Int'l Legal Materials* 685 (1963). The Nuclear Operators Convention and the other three are reprinted in *The International Law of Pollution* 433, 445, 422, 439 (J. Barros and D. Johnston eds 1974). Other conventions on international protection against radiation hazards appear in ibid at 404–64; 2 *New Directions in the Law of the Sea*, supra note 29, at 792–4.

63 *Done* 29 November 1969, [1975] 1 U.S.T. 765, T.I.A.S. no. 8068

64 *Done* 29 November 1969, in 9 *Int'l Legal Materials* 45 (1970). See also Protocol to the International Convention on Civil Liability for Oil Pollution Damage, 19 November 1976, ibid, vol. 16, at 607 (1977); the protocol incorporates the new unit of account, SDRs, supra note 61.

65 Ibid, vol. 9, at 66

66 Ibid, at 67

67 Ibid, at 65

68 Supra note 63, art. 1

69 In fact, the convention may be even more restrictive than the traditional rights (see 46–7 infra), since it requires various forms of notification and consultation before a state takes measures (supra note 63, art. 3).

70 Supra note 64, art. 2

71 *Done* 18 December 1971, in 11 *Int'l Legal Materials* 284, art. 3 (1972). See also Protocol to the International Convention on the Establishment of an International Fund for Compensation for Oil Pollution Damage, 19 November 1976, ibid, vol. 16, at 621 (1977); the protocol incorporates SDRs, supra note 61.

72 *Done* 9 June 1969, ibid, vol. 9, at 359 (1970)

73 Ibid, art. 1 (emphasis added)

74 The zones are described by latitude and longitude in the Annex to the Agreement. There are some zones of joint responsibility.
75 Arctic Waters Pollution Prevention Act of 1970, supra note 43; Oil in Navigable Waters Act, c. 21 (U.K.). For discussion, see 27 supra.
76 *Report of the Second Session of the Inter-Governmental Working Group on Marine Pollution* 7–8 and 8–12, U.N. Doc. A/Conf.48/IWGMP.II/5 (1971), reprinted in *Report*, supra note 3, at 48 and Annex 3, at 1–3.
77 Statement of Objectives, ibid. The full text reads: 'The marine environment and all the living organisms which it supports are of vital importance to humanity, and all people have an interest in assuring that this environment is so managed that its quality and resources are not impaired. This applies especially to coastal nations, which have a particular interest in the management of coastal area resources. The capacity of the sea to assimilate wastes and render them harmless, and its ability to regenerate natural resources, is not unlimited. Proper management is required and measures to prevent and control marine pollution must be regarded as an essential element in this management of the oceans and seas and their natural resources.' Then follow the twenty-three General Principles.
78 Ibid, at 12–13
79 *Done* 15 February 1972, in 11 *Int'l Legal Materials* 262 (1972). See Yturriaga-Barberan, Convenio de Oslo de 1972 para le prevención de la contaminación porvertidos, 1 *Revista de Instituciónes Europeas* 121 (1974).
80 The provisions are contained in arts. 5 and 6 and the lists of substances in Annexes I and II of the Convention, supra note 79.
81 For example, article 8 of the Oslo Convention contained the quite straightforward exception that its provisions 'shall not apply in case of force majeure due to stress of weather or any other cause when the safety of human life or of a ship or aircraft is threatened'; and, in addition, article 9 offered a more obtuse escape clause: 'If a Contracting Party in an emergency considers that a substance listed in Annex I to this Convention [prohibited substances and materials] cannot be disposed of on land without unacceptable danger or damage, the Contracting Party concerned will forthwith consult the Commission. The Commission shall recommend methods of storage or the most satisfactory means of destruction or disposal under the prevailing circumstances. The Contracting Party shall inform the Commission of the steps adopted in pursuance of its recommendation. The Contracting Parties pledge themselves to assist one another in such situations.' The type of situation envisaged in article 9 is said already to have occurred in Operation CHASE, when US authorities had urgently to dispose of nerve gas cannisters in a dangerous condition. For discussion of the unclear legal requirements, see Brown, The Conventional Law of the Environment, 13 *Natural Resources J.* 203, 228 (1973), and

his The Ocean Dumping of Nerve Gas: A Case Study of 'Operation CHASE,' ibid, vol. 11, at 249 (1971).

82 Oslo Anti-Dumping Convention, supra note 79, art. 7 and Annex III
83 Ibid, art. 15
84 Recommendation 92, *Report*, supra note 3, at 48
85 Recommendation 86, ibid, at 45
86 Ibid, at 4 and 45-9
87 G.A. Res. 2994, 27 U.N. GAOR Supp. 30, at 42, U.N. Doc. A/8730 (1972). See also related G.A. Res. 2995-7, ibid, at 42-3.
88 *Done* 29 December 1972, [1975] 2 U.S.T. 2403, T.I.A.S. no. 8165. For discussion see de Mestral, La convention sur la prévention de la pollution résultant de l'immersion de déchets, 11 *Can. Y.B. Int'l L.* 226 (1973).
89 The prohibitions and a permit system for restricted substances are provided for in art. 4, and the lists themselves are contained in Annexes I and II to the Convention, supra note 88.
90 Ibid, art. 7
91 Ibid, art. 13
92 *Done* 2 November 1973, in 12 *Int'l Legal Materials* 1319 (1973)
93 Ibid, preamble
94 Ibid, Annex 1. In this annex, a 'special area' is defined as 'a sea area where for recognized technical reasons in relation to its oceanographical and ecological condition and to the particular character of its traffic the adoption of special mandatory methods for the prevention of sea pollution by oil is required' (Reg. 1(10)). For its purposes, there are five internationally agreed-upon special areas: the Mediterranean Sea area, the Baltic Sea area, the Black Sea area, the Red Sea area, and the 'Gulfs area' (ibid, Reg. 10).
95 See 35-6 and note 79 supra.
96 *Done* 4 June 1974, in 13 *Int'l Legal Materials* 352 (1974)
97 Ibid, art. 3
98 Supra note 96, art. 12
99 Supra note 79, arts. 16-18
100 Supra note 96, arts. 15-16
101 Res. no. 2, in 13 *Int'l Legal Materials* 373 (1974)
102 *Done* 1 November 1974, in International Conference on Safety of Life at Sea: Final Act of the Conference Including the Convention, in 14 *Int'l Legal Materials* 959 (1975)
103 Convention for the Protection of the Mediterranean Sea against Pollution (Barcelona Convention), *done* 16 February 1976, in 15 *Int'l Legal Materials* 290 (1976); Protocol for the Prevention of Pollution of the Mediterranean Sea by

Dumping from Ships and Aircraft, ibid, at 300; Protocol concerning Co-operation in Combatting Pollution of the Mediterranean Sea by Oil and Other Harmful Substances in Cases of Emergency, ibid, at 306. The Mediterranean Convention and its protocols entered into force on 12 February 1978. Meanwhile, the states concerned are now working on another protocol on land-based sources of pollution. See UNEP Report on Draft Protocol for the Protection of the Mediterranean Sea against Pollution from Land-Based Sources, in ibid, vol. 16, at 958 (1977).

In May 1978, the eight countries surrounding the Persian-Arabic Gulfs managed to agree on a similar framework convention and two protocols, also under the auspices of UNEP. See Bahrain-Iran-Iraq- Kuwait-Oman-Quatar-Saudi Arabia-United Arab Emirates: Agreements from the Kuwait Regional Conference on the Protection and Development of the Marine Environment and the Coastal Areas, 17 *Int'l Legal Materials* 501 (1978).

104 G.A. Res. 1962, supra note 16. See generally Taubenfeld, International Environmental Law: Air and Outer Space, 13 *Natural Resources J.* 315 (1973).
105 Supra note 16
106 Ibid, art. 6
107 Ibid, art. 7
108 Ibid, art. 4
109 Ibid, art. 9
110 Ibid, art. 8. The return provision was subsequently expanded upon by the Agreement on the Rescue of Astronauts, the Return of Astronauts and the Return of Objects Launched into Outer Space, *done* 22 April 1968, [1968] 6 U.S.T. 7570, T.I.A.S. no. 6599, 672 U.N.T.S. 119. Article 5 of the latter agreement, however, contains an exception to recovery and return provisions when a space object or its component parts 'is of a hazardous or deleterious nature.' On registry, see also Convention on the Registration of Objects Launched into Outer Space, *done* 14 January 1975, T.I.A.S. no. 8480.
111 G.A. Res. 2777, 26 U.N. GAOR Supp. 29, at 25, U.N. Doc. A/8429 (1971)
112 *Done* 29 March 1972, [1973] 2 U.S.T. 2389, T.I.A.S. no. 7762. See also art. 4, which spells out the same double standard of liability for injury to interests of third parties on the surface of the earth and elsewhere.
113 Ibid, art. 8
114 Ibid, art. 12
115 The 'Partial Test Ban' Treaty, *done* 5 August 1963, [1963] 2 U.S.T. 1313, T.I.A.S. no. 5433, 480 U.N.T.S. 43
116 Ibid, art. 1
117 Respectively, *done* 1 July 1968, [1970] 1 U.S.T. 483, T.I.A.S. no. 6839, 729 U.N.T.S. 161; *done* 11 February 1971, [1973] 1 U.S.T. 701, T.I.A.S. no. 7337; and *done* 14

February 1967, 634 U.N.T.S. 281. By the 'Seabed Denuclearization' Treaty, which entered into force 18 May 1972, the states parties, *inter alia*, undertake 'not to emplant or emplace on the seabed and the ocean floor and in the subsoil thereof beyond the outer limit of a seabed zone ... any nuclear weapons or any other types of weapons of mass destruction as well as structures, launching installations or any other facilities specifically designed for storing, testing or using such weapons' (ibid, art. 1). As regards the Latin American treaty, see the U.N. General Assembly resolutions on the Tlatelolco Treaty: G.A. Res. 2830, 26 U.N. GAOR Supp. 29, at 35, U.N. Doc. A/8429 (1971); and the two earlier resolutions cited therein; see also Additional Protocol II to the Treaty of 14 February 1967 for the Prohibition of Nuclear Weapons in Latin America, *done* 14 February 1967, [1971] 1 U.S.T. 754, T.I.A.S. no. 7137, 634 U.N.T.S. 364.

118 *Done* 10 April 1972, [1975] 1 U.S.T. 583, T.I.A.S. no. 8062

119 *Report*, supra note 3, at 66

120 *Nuclear Tests (Australia* v *France)* and *Nuclear Tests (New Zealand* v *France)*, [1974] I.C.J. 253 and 457

121 Interim Protection, Orders of 22 June 1973, [1973] I.C.J. 99 and 135

122 Judgments of 20 December 1974, [1974] I.C.J. 253 and 457

123 Applications to Intervene, Orders of 20 December 1974, [1974] I.C.J. 530 and 535

124 1 December 1959, [1961] 1 U.S.T. 794, T.I.A.S. no. 4780, 402 U.N.T.S. 71. See also Convention for the Conservation of Antarctic Seals, 11 February 1972, in 11 *Int'l Legal Materials* 251 (1972). For discussion see Hambro, Some Notes on the Future of the Antarctic Treaty Collaboration, 68 *Am. J. Int'l L.* 217 (1971); Hayton, The Antarctic Settlement of 1959, ibid, vol. 54, at 349 (1960); Taubenfeld, A Treaty for Antarctica, 1961 *Int'l Conc.* no. 531; Note, Natural Resources Jurisdiction on the Antarctic Continental Margin, 11 *Va. J. Int'l L.* 374 (1971).

125 Antarctic Treaty, supra note 124, art. 4

126 Ibid, art. 5

127 Ibid, art. 1

128 Ibid, art. 9

129 For discussion of the environmental implications of the present negotiations, see Mink, Oceans: Antarctic Resources and Environmental Concerns, 78 *Dept. State Bull.* 2013, at 51 (April 1978). There are now nineteen parties to the Antarctic Treaty; besides the 'Thirteen' consultative parties named in the text, they comprise Brazil, Czechoslavakia, the German Democratic Republic, the Netherlands, Romania, and South Africa. Other countries which have joined in activities in the area include the Federal Republic of Germany and South Korea.

130 See note 17 supra.

131 11 January 1909, 36 Stat. 2448 (1909–11), T.S. no. 548. See also Convention to Regulate the Level of the Lake of the Woods, 24 February 1925, 44 Stat. 2108 (1925–7), T.S. no. 721, 43 L.N.T.S. 251; Convention Providing for Emergency Regulation of Rainy Lake and of Certain Other Boundary Waters, 15 September 1938, 54 Stat. 1800 (1939–41), T.S. no. 961, 203 L.N.T.S. 267; Agreement Relating to the Upper Columbia River Basin, 3 March 1944, 58 Stat. 1236 (1944), E.A.S. 399, 109 U.N.T.S. 191; Treaty Relating to Uses of the Niagara River, 27 February 1950, [1950] 1 U.S.T. 694, T.I.A.S. no. 2130, 132 U.N.T.S. 223. The United States and Canada have signed and had entered into force several other treaties in relation to boundary waters, particularly in connection with the construction and operation of the St Lawrence Seaway.

132 U.S.-Canada Boundary Waters Treaty, supra note 131, art. 4

133 See IJC *Report on Pollution of Lake Erie, Lake Ontario and the International Sections of the St Lawrence River* 2 (1970). See also 3 M. Whiteman, supra note 7, at 828.

134 IJC *Report*, supra note 133. See also IJC *Report on the Pollution of Rainy River and Lake of the Woods* (1965).

135 15 April 1972, [1972] 1 U.S.T. 301, T.I.A.S. no. 7312. See generally Bilder, Controlling Great Lakes Pollution: A Study in U.S.-Canadian Environmental Cooperation, in *Law, Institutions and the Global Environment* 294 (J. Hargrove ed 1972).

136 See Meyers, The Colorado Basin, in *The Law of International Drainage Basins* (A. Garretson, R. Hayton, and C. Olmstead eds 1967). See also Brownell and Eaton, Colorado River Salinity Problem with Mexico, 69 *Am J. Int'l L.* 255 (1975); Weinberg, 'Salt Talks' United States and Mexican Style: A Case Study of the Lower Colorado Salinity Dispute, in *International Responsibility for Environmental Injury*, supra note 7; 3 M. Whiteman, supra note 7, at 945–66. See generally, Pollution and Political Boundaries: U.S.-Mexican Environmental Problems, 12 *Natural Resources J.* no. 4 (1972).

137 Colorado River Compact of 1922. The details of its negotiation and ratification and of congressional consent, as well as commentary on its provisions, are set forth in Meyers, supra note 136, at 504–11.

138 Treaty Relating to the Utilization of Waters of the Colorado and Tijuana Rivers and of the Rio Grande, 14 November 1944, 59 Stat. 1219 (1945), T.S. no. 994, 3 U.N.T.S. 313

139 Ibid, art. 10

140 For discussion, see Weinberg, supra note 136.

141 *Done* 30 August 1973, [1973] 2 U.S.T. 1968, T.I.A.S. no. 7780. For an earlier five-year temporary measure, see Minute 218 of the International Boundary

and Water Commission, 22 March 1965, and Agreement Extending the Provisions of Minute 218, *entered into force* 16 November 1970, [1970] 3 U.S.T. 2478, T.I.A.S. no. 6988. See also Minute no. 241 of the International Boundary and Water Commission to Improve Immediately the Quality of Colorado River Waters Going to Mexico, *entered into force* 14 July 1972, T.I.A.S. no. 7404.

142 Colorado River Salinity Agreement, supra note 141, art. 1

143 Report of the Committee on the Uses of Waters of International Rivers, in ILA, *Report of the Fifty-Second Conference Held at Helsinki* 478 (1967)

144 Draft Convention, in Ad Hoc Committee of Experts to Prepare a European Convention on the Protection of International Fresh Waters Against Pollution, *Report*, Council of Europe Doc. EXP/Eau (74) 6 Add. 1 (13 March 1974). Three years later, on 31 March 1977, a written question was submitted to the Parliamentary Assembly of the Council of Europe asking about any changes to the draft convention and its adoption and possible opening for signature (Parliamentary Assembly, Council of Europe, 29th Ordinary Session, 2 *Documents* 1, Doc. no. 3961 (25–9 April 1977)). The draft still remains in limbo.

145 Draft Explanatory Report on the European Convention for the Protection of International Watercourses Against Pollution, in ibid, Add. 2, at 27

146 See generally Utton, International Water Quality Law, 13 *Natural Resources J.* 282 (1973). Institutions, however, have been slower to develop. For example, the Central Commission for the Navigation of the Rhine was established in 1815, but the International Commission for the Protection of the Rhine Against Pollution did not come into being until 1963; and the situation of the Danube is said to be even worse, with no body specially concerned with its pollution and little likelihood that the Danube Commission will expand its role sufficiently to uphold such common interests in environmental protection of the river. See Stein, The Potential of Regional Organizations in Managing Man's Environment, in *Law, Institutions and the Global Environment*, supra note 135, at 253, 265–70. For some recent developments, see Convention on the Protection of the Rhine against Chemical Pollution, *done* 3 December 1976, in 16 *Int'l Legal Materials* 242 (1977); Convention on Protection of the Rhine against Pollution by Chlorides, *done* 3 December 1976, in ibid, at 265.

147 See Recommendation 55, in *Report*, supra note 3, at 35–6.

148 Supra note 96, arts. 3 and 4

149 ICNT, art. 208(1), in 8 *Third UNCLOS: Official Records*, supra note 2, at 36. For discussion, see 26 supra.

150 See 34 and note 63 supra.

151 See *U.N. Charter* art. 51 and the complementary provision prohibiting 'the threat or use of force' in ibid, art. 2 para. 4. On the definition and differentia-

tion of the various forms of self-help, see M. McDougal and F. Feliciano, *Law and Minimum World Public Order* 212–16 *et seq.* (1961).

152 See Brown, The Lessons of the *Torrey Canyon*, 21 *Current Legal Prob.* 113 (1968); Utton, Protective Measures and the Torrey Canyon, 9 *B.C. Ind. & Com. L. Rev.* 613 (1968). For a more expansive view of international law doctrines of protection, see Hardy, International Protection against Nuclear Risks, 10 *Int'l & Comp. L.Q.* 739 (1961); McDougal and Schlei, The Hydrogen Bomb Tests in Perspective: Lawful Measures for Security, in M. McDougal and Associates, *Studies in World Public Order* 273 (1960).

153 See generally E. Cowan, *Oil and Water: The Torrey Canyon Disaster* 193–6 and passim (1968).

154 For discussion of the increased risks due to developments in transportation, see Goldie, Development of an International Environmental Law – An Appraisal, in *Law, Institutions and the Global Environment*, supra note 135, at 104, 125–6. Deepwater ports, as their name indicates, are facilities that are being developed to enable oil tankers to unload offshore without having to enter traditional ports. See also Mostert, Profiles: Supertankers, 50 *New Yorker* 45 (13 May 1974) and 46 (20 May 1974); N. Mostert, *Supership* (1974).

155 See 34 and note 63 supra. See also ICNT art. 222, in 8 *Third UNCLOS: Official Records*, supra note 2, at 39.

156 See *International Responsibility for Environmental Injury*, supra note 7. See also Beesley, The Canadian Approach – Environmental Law on the International Plane, in *Private Investors Abroad – Problems and Solutions in International Business* 239 (Proc. Southwestern Legal Fnda. 1973); Bleicher, An Overview of Environmental Legislation, 2 *Ecology L.Q.* 1 (1972); Olmstead, Prospects for Regulation of Environmental Conservation under International Law, in ILA *The Present State of International Law* 245 (1973).

157 See 21 supra (emphasis added).

158 Supra note 87

159 See 'Facilitating constructive use' 53–66 infra.

160 *Report*, supra note 3, at 7

161 (*United States* v *Canada*), 3 U.N.R.I.A.A. 1911 and 1938 (1938 and 1941), reprinted in 33 *Am. J. Int'l L.* 182 (1939) and ibid, vol. 35, at 684 (1941)

162 [1949] I.C.J. 4

163 (*Spain* v *France*), 12 U.N.R.I.A.A. 281 (1957), digested in 53 *Am. J. Int'l L.* 156 (1959)

164 See Canada-United States Settlement of Gut Dam Claims, 22 September 1968, Report of the Agent of the United States before the Lake Ontario Claims Tribunal, in 8 *Int'l Legal Materials* 118 (1969). The *compromis* establishing the

tribunal (25 May 1965, T.I.A.S. no. 6114), a map indicating the location of Gut Dam, and a report on the US Foreign Claims Settlement Commission and the Lake Ontario Claims Program appear respectively in ibid, vol. 4, at 468, 472, and 473 (1965). See also Agreement on Settlement of Claims Relating to Gut Dam, 18 November 1968, [1968] 6 U.S.T. 7863, T.I.A.S. no. 6624.

165 L.F.E. Goldie has written at length on this subject. See, eg, Goldie, International Principles of Liability for Pollution, 9 *Colum. J. Transnat'l L.* 283 (1970).

166 Supra note 161

167 This principle has been especially extensively discussed in the forum of the Organization for Economic Co-operation and Development. See OECD, Guiding Principles on the Environment, 11 *Int'l Legal Materials* 1172 (1972), reprinted in ibid, vol. 14, at 236 (1975). See also OECD, *Society and the Assessment of Technology* (1973); OECD, *Problems in Transfrontier Pollution* (1974). The 'polluter pays' principle and related matters will be discussed in chapter 4, 103–4. See also 155–8. .

168 3 U.N.R.I.A.A. 1963

169 Ibid, at 1965–6

170 Supra note 162

171 [1949] I.C.J. 22

172 Supra note 163

173 For consideration of this significance, see Laylin and Bianchi, The Role of International Adjudication in International River Disputes: The Lake Lanoux Case, 53 *Am. J. Int'l L.* 30 (1959).

174 12 U.N.R.I.A.A. 303

175 Supra note 164

176 Decision of 15 January 1968, Transcript at 589, quoted in *Report*, supra note 164, 8 *Int'l Legal Materials* 136. It is worth noting, however, that these responsibilities are reciprocal. Canada, on its part, has been disturbed by the effects of the proposed Garrison Diversion, since certain uses of the waters in North Dakota would threaten injury to interests in Manitoba. There have been several diplomatic exchanges, and the matter has been considered by the International Joint Commission, which recommended that this US project not be carried out at this time. See IJC, Transboundary Implications of the Garrison Diversion Unit (1977). For discussion, see chapter 6, 166.

177 See 46–7 supra.

178 Draft Declaration on the Human Environment, U.N. Doc. A/Conf. 48/4, Annex, para. 20, at 4 (1972). For exposition of the history of 'draft Principle 20,' see Sohn, supra note 5, at 496–502.

179 Brazil was undertaking feasibility studies for the construction of a giant hydro-electric installation on the Parana River, which eventually flows into Argentina,

becoming the La Plata River. Argentina was concerned that the power project would cause alteration of the river, resulting in floods, droughts, and water pollution injury to Argentinian interests. Argentina therefore requested of Brazil that consultations be held before construction was undertaken, which request was rejected. As a result, these two countries were unable to agree on the content of 'draft Principle 20' in the intergovernmental Working Group on the Declaration, and they continued to disagree at the Stockholm Conference itself. Brazil's opposition to a strict consultation requirement was supported by many developing countries, who were afraid such a duty would cause impediments and delays in development programs. Consequently, no accord was reached at Stockholm, and the matter was left for the General Assembly.

180 *Report*, supra note 3, at 119
181 Co-operation between States in the Field of the Environment, G.A. Res. 2995, supra note 87
182 G.A. Res. 3129, 28 U.N. GAOR Supp. 30, at 48, 49, U.N. Doc. A/9030 (1973). For UNEP's response, see Co-operation in the Field of the Environment Concerning Natural Resources Shared by Two or More States: Report of the Executive Director, U.N. Doc. UNEP/GC/44 (1975). The Executive Director subsequently established an Intergovernmental Working Group of Experts on Natural Resources Shared by Two or More States. For the latest report of this group, see U.N. Doc. UNEP/IG.12/2 (1978).
183 OECD Doc. C(74)224 (21 November 1974), reprinted in 14 *Int'l Legal Materials* 242 (1975). See also Council Recommendation on Implementing a Regime of Equal Right of Access and Non-Discrimination in relation to Transfrontier Pollution, 17 May 1977, ibid, vol. 16, at 977 (1977).
184 Ibid, at 246, Title F
185 Ibid, Title E
186 *Report*, supra note 3, at 40. See 29 supra.
187 Supra note 58, art. 2
188 Ibid, art. 4
189 Ibid, art. 5
190 Ibid, art. 7
191 42 U.S.C. § 4321, 4332(2)(C) and (E) (1970). See *National Organization for the Reform of Marijuana Laws* (NORML) v *Department of State*, 452 F.Supp. 1226 (D.D.C. 1978); *Sierra Club* v *Coleman*, 421 F.Supp. 63 (D.D.C. 1976); *Environmental Defense Fund, Inc.* v *Agency for International Development*, 6 ELR 20121 (D.D.C. 1975); *Sierra Club* v *Atomic Energy Commission*, 4 ELR 20685 (D.D.C. 1974). See also Forthcoming CEQ Regulations to Determine Whether NEPA Applies to Environmental Impacts Limited to Foreign Countries, 8 ELR 10111 (1978).
192 ICNT art. 207, in 8 *Third UNCLOS: Official Records*, supra note 2, at 36

193 Trans-Alaska Pipeline Authorization Act, 43 U.S.C.A. § 1651, 1653(c) (1) (Supp. 1975) (emphasis added)

194 *Wilderness Soc'y* v *Morton*, 463 F.2d 1261 (D.C. Cir. 1972). For the substantive issues, see *Wilderness Soc'y* v *Morton*, 479 F.2d 842 (D.C. Cir. 1972). The case was brought by various US environmental groups to halt construction of the Trans-Alaska Pipeline. A Canadian citizen and a Canadian environmental group were allowed to intervene upon a finding by the Court that their interests, as compared to those of the plaintiffs, were 'sufficiently antagonistic in this litigation to require granting of the application for intervention' (463 F.2d at 1262–3). See also *Wilderness Soc'y* v *Hickel*, 325 F. Supp. 422 (D.D.C. 1970).

195 Convention on the Protection of the Environment, *done* 19 February 1974, in 13 *Int'l Legal Materials* 591 (1974). Denmark, Finland, Norway, and Sweden are the parties to this treaty.

196 But at least one scholar claims that this is a 'human chauvinistic' and incomplete legal objective (Stone, Should Trees Have Standing? Toward Legal Rights for Natural Objects, 45 *S. Cal. L. Rev.* 450 (1972)).

197 Principle 1, *Report*, supra note 3, at 4 (emphasis added)

198 The Food and Agriculture Organization Department of Fisheries circulated during the Geneva session of UNCLOS III an informal paper on Definitions of Some Concepts in Fishery Management (1975), which also favoured the concept of optimum sustainable yield over maximum yield. On the 'meaning of conservation,' see S. von Ciriacy-Wantrup, supra note 19, at 48–61. See also ICNT art. 161(3), in 8 *Third UNCLOS: Official Records*, supra note 2, at 14.

199 For discussion, see M. McDougal and W. Burke, supra note 10, at 469–82.

200 5 April 1956, 231 U.N.T.S. 199, 1956 U.K.T.S. 8; amended in 1958, 1960, 1961, 1962, and 1963. See 2 *New Directions in the Law of the Sea*, supra note 29, at 782–3.

201 24 January 1959, 486 U.N.T.S. 157, 1963 U.K.T.S. 68, art. 7

202 2 *New Directions in the Law of the Sea*, supra note 29, at 782–9. For the United States as an illustration, see Jacobs, United States Participation in International Fisheries Agreements, 6 *J. Maritime L. & Com.* 471 (1975).

203 Supra note 60

204 Ibid, art. 6

205 Ibid, art. 7 (again subject to mandatory dispute settlement provisions)

206 7 July 1911, 37 Stat. 1542 (1911–13), T.S. no. 564

207 More specifically, under the convention, ibid, in return for refraining from sealing (an otherwise perfectly lawful activity according to prior arbitration decisions), Canada and Japan were to receive percentages of the gross catch of the United States and Russia, as well as lump-sum payments from the United States. Similarly, Japan was to provide a share of its catch to the other three

states, and if seals were caught within British jurisdiction (in Canadian areas) similar sharing was also to apply.

208 9 February 1957, [1957] 2 U.S.T. 2283, T.I.A.S. no. 3948, 314 U.N.T.S. 105; Amendment and extension 8 October 1963, [1964] 1 U.S.T. 316, T.I.A.S. no. 5558, 494 U.N.T.S. 303; Amendment and extension 3 September 1969, [1969] 3 U.S.T. 2992, T.I.A.S. no. 6774; Amendment and extension 7 May 1976, T.I.A.S. no. 8368

209 Interim Convention, supra note 208, preamble. For discussion, see M. McDougal and W. Burke, supra note 10, at 942–3.

210 Interim Convention, supra note 208, art. 2 states the parties' agreement to cooperate in determining both the measures necessary to make possible the maximum sustained productivity of the seals and also 'what the relationship is between fur seals and other living marine resources and whether fur seals have detrimental effects on other living marine resources exploited by any of the Parties and, if so, to what extent.'

211 *Done* 8 February 1949, [1950] 1 U.S.T. 477, T.I.A.S. no. 2089, 157 U.N.T.S. 157. Regarding this convention, see Protocol, *done* 25 June 1956, [1959] 1 U.S.T. 59, T.I.A.S. no. 4170, 331 U.N.T.S. 388; Declaration of Understanding, *done* 24 April 1961, [1963] 1 U.S.T. 924, T.I.A.S. no. 5380, 480 U.N.T.S. 334; Protocol, *done* 15 July 1963, [1966] 1 U.S.T. 635, T.I.A.S. no. 6011, 590 U.N.T.S. 292; Protocol, *done* 29 November 1965, [1969] 1 U.S.T. 567, T.I.A.S. no. 6840; Protocol, *done* 29 November 1965, [1970] 1 U.S.T. 576, T.I.A.S. no. 6841; Protocol, *done* 1 October 1969, [1972] 2 U.S.T. 1504, T.I.A.S. no. 7432.

212 The original parties were Canada, Denmark, France, Iceland, Italy, Newfoundland, Norway, Portugal, Spain, the United Kingdom, and the United States. The following additional states have since become parties: Bulgaria, the Federal Republic of Germany, Japan, Poland, Romania, and the USSR; meanwhile, of course, Newfoundland has joined Canada.

213 For discussion of ICNAF as a regulatory regime, see D. Johnston, *The International Law of Fisheries* 365–9 (1965). Also discussed are commissions and other arrangements for Northeast Atlantic, Northeast Pacific, Northwest Pacific, and Antarctic living resources. On regional fisheries organizations generally, see A. Koers, *International Regulation of Marine Fisheries* (1973).

214 Convention for the Regulation of Whaling, 24 September 1931, 49 Stat. 3079 (1935–6), T.S. no. 880, 155 L.N.T.S. 349; International Agreement for the Regulation of Whaling, 8 June 1937, 52 Stat. 1460 (1938), T.S. no. 993, 190 L.N.T.S. 80

215 2 December 1946, 62 Stat. 1716 (1948), T.I.A.S. no. 1849, 161 U.N.T.S. 72, with twenty-five amendments through 1973

216 Ibid, art. 3

217 Recommendation 33, *Report*, supra note 3, at 23. See also *Reports of the Governing Council of the United Nations Environment Programme* 28 U.N.

GAOR Supp. 25, at 10, U.N. Doc. A/9025 (1973); 29 U.N. GAOR Supp. 25, at 149, U.N. Doc. A/9625 (1974); 30 U.N. GAOR Supp. 25, at 112, U.N. Doc. A/10025 (1975) [1st, 2d, and 3d sess. respectively]. On the controversy, see Efforts to Stop Whaling Stir Backlash in Japan, *N.Y. Times*, 22 June 1974, at 10, col. 3; but see Whaling Meeting Shows Signs of a Priority for Conservation, ibid, 1 July 1974, at 7, col. 1.

218 ICNT art. 65, in 8 *Third UNCLOS: Official Records*, supra note 2, at 15

219 *Done* 13 September 1973, in 12 *Int'l Legal Materials* 1291 (1973) and *done* 22 March 1974, ibid, vol. 13, at 546 (1974)

220 *Done* 25 February 1972, in 1 *New Directions in the Law of the Sea*, supra note 29, at 449

221 Agreement on the Regulation of the Fishing of North-East Arctic (Arcto-Norwegian) Cod, *done* 15 March 1974, in 13 *Int'l Legal Materials* 1261 (1974)

222 11 December 1970, [1970] 3 U.S.T. 2664, T.I.A.S. no. 7009, with Protocol of the same date, [1971] 1 U.S.T. 113, T.I.A.S. no. 7043, preamble. This treaty has since been replaced by another agreement of the same name and with the same expressed purposes, 21 June 1973, [1973] 3 U.S.T. 1603, T.I.A.S. no. 7664. The United States and the USSR have signed several bilateral fisheries agreements in the past few years, including four on 21 February 1973: Agreement Relating to Fishing for King and Tanner Crab, Agreement Relating to Fishing Operations in the North-eastern Pacific Ocean, Agreement on Certain Fisheries Problems in the North-eastern Part of the Pacific Ocean Off the Coast of the United States, and Agreement Relating to the Consideration of Claims Resulting from Damage to Fishing Vessels or Gear and Measures to Prevent Fishing Conflicts, [1973] 2 U.S.T. 603, 617, 631, and 669, T.I.A.S. nos. 7571, 7572, 7573, and 7575, with a Protocol to the last of 21 June 1973, [1973] 2 U.S.T. 1588, T.I.A.S. no. 7663.

223 Agreement Concerning Shrimp, 9 May 1972, [1973] 1 U.S.T. 923, T.I.A.S. no. 7603, replaced upon expiration by Agreement Concerning Shrimp, *done* 14 March 1975, in 14 *Int'l Legal Materials* 910 (1975), and Agreement Concerning Shrimp, 19 May 1972, in 1 *New Directions in the Law of the Sea*, supra note 29, at 463

224 *Done* 15 November 1973, in 13 *Int'l Legal Materials* 13 (1974)

225 Supra note 124. The new agreement recalls and extends the Agreed Measures for the Conservation of Antarctic Fauna and Flora adopted under the Antarctic Treaty of 1959 as regards the seals (ibid, preamble).

226 Convention on International Trade in Endangered Species of Wild Flora and Fauna, *done* 3 March 1973, [1976] 2 U.S.T. 1087, T.I.A.S. no. 8247.

227 See 28 and 24–5 supra.

228 *Fisheries Jurisdiction (United Kingdom of Great Britain and Northern Ireland* v *Iceland)* and *Fisheries Jurisdiction (Federal Republic of Germany* v *Iceland)*, [1974] I.C.J. 3 and 175

229 Agreement Settling the Fisheries Dispute, 11 March 1961, 397 U.N.T.S. 275
(1961), and Agreement Concerning the Fishery Zone Around Iceland, 19 July
1961, 409 U.N.T.S. 47 (1961). The former agreement, between the United Kingdom and Iceland, is reprinted in 11 *Int'l Legal Materials* 490 (1972). Cf
Belgium-Iceland: Fisheries Agreement on the Extension of the Icelandic
Fishery Limits to 200 Miles, 28 November 1975, ibid, vol. 15, at 1 (1976).
Federal Republic of Germany-Iceland: Fisheries Agreement on the Extension of the Icelandic Fishery Limits to 200 Miles, 28 November 1975, ibid, at
43; Iceland-Norway: Agreement concerning Norwegian Fishing in Icelandic
Waters 10 March 1975, ibid, at 875; Iceland-United Kingdom: Agreement
concerning British Fishing in Icelandic Waters, 1 June 1976, ibid, at 878.

230 15 February 1972, in 11 *Int'l Legal Materials* 643, paras. 2, 2, and 4
respectively (1972)

231 14 July 1972, in ibid at 1112, art. 2

232 See Interim Protection, Orders of 17 August 1972, [1972] I.C.J. 12 and 30; In-
terim Measures, Orders of 12 July 1973, [1973] I.C.J. 302 and 313; Jurisdiction of
Court, Judgments, [1973] I.C.J. 3 and 49. These are reprinted in 11 *Int'l Legal
Materials* 1079 (1972); ibid vol. 12, at 743 (1973); ibid, at 290.

233 Merits, Judgment, [1974] I.C.J. 3, 34–5 and 175, 205–6

234 [1974] I.C.J. 23–4

235 See, eg, Summary of Report by the Secretary-General entitled *Possible Impact
of Seabed Mineral Production in the Area Beyond National Jurisdiction on
World Markets, with Special Reference to the Problems of Developing Countries: A Preliminary Assessment* (U.N. Doc. A/AC.138/36), in *Report of the
Committee on the Peaceful Uses of the Sea-Bed and the Ocean Floor Beyond
the Limits of National Jurisdiction* 26 U.N. GAOR Supp. 21, at 227, U.N. Doc.
A/8421 (1971); Additional Notes on the Possible Economic Implications of
Mineral Production from the International Sea-Bed Area: Report of the Sec-
retary-General, *Seabeds Report*, supra note 33, at 109. See also Distribution
of Manganese Nodules on the Deep Ocean floor, presented by V.E.
McKelvrey, US Delegation, at the 27 April 1978 Seminar Sponsored by
Negotiating Group 1 of UNCLOS III (2 May 1978).

236 See Organization of the Work of the Seventh Session, U.N. Doc. A/Conf.62/62 (13
April 1978).

237 Since there was insufficient time at the Geneva portion of the seventh session of
UNCLOS III, and since it was decided to reconvene the same session in New York
from 21 August to 15 September 1978, no revision of the ICNT, supra note 2, was
attempted in Geneva. The progress achieved there was, however, recorded and
asembled in *Reports of the Committees and Negotiating Groups on Negotiations at the Seventh Session contained in a Single Document Both for Pur-*

poses of Record and for the Convenience of Delegations (19 May 1978); the
reports of Negotiating Groups 1 to 3, dealing with First Committee matters, are
found at pages 1–70 of this compilation, which does not bear a standard UN
document number.

238 See note 16 supra.

239 Ibid, preamble to both. See also 39 supra.

240 The Outer Space Treaty, ibid, art. 2

241 For definitions see note 19 supra.

242 A merchant vessel on the high seas or a ship of war anywhere is assimilated to the
territory of the state in which it is registered; that state thus retains jurisdiction
over it. The jurisdiction of a state over its vessels in foreign national waters is
concurrent with that of the coastal state (with certain qualifications, such as distinctions between civil and criminal questions and exceptions for 'distress' and
'innocent passage'). Neither in shared nor in national waters, therefore, does a
state of registry entirely lose control over its vessels. The major exception to this
set of jurisdictional principles, however, is that all states are authorized to take
measures against piracy, irrespective of the national character of the offending
ship. On public order and the nationality of ships, see M. McDougal and W. Burke,
supra note 10, at 1008–140.

243 The International Air Services Transit Agreement (the 'Two Freedoms'),
opened for signature 7 December 1944, 59 Stat. 1963 (1945), E.A.S. no. 487, 84
U.N.T.S. 389, art. 1, § 5, for example, expresses the jurisdictional principles
governing aircraft in a negative fashion: it allows any state to revoke a cer-
tificate or permit 'in any case where it is not satisfied that substantial ownership and effective control are vested in the nationals of a contracting State.'

244 The Outer Space Treaty, supra note 16, art. 8, provides that a state on whose
registry an object launched into space is carried 'shall retain jurisdiction and
control over such object and over any personnel thereof, while in outer space or
on a celestial body.' When a space object returns to earth in any place not under
the jurisdiction of the launching state, the Rescue and Return Agreement, supra
note 110, art. 5, requires any state receiving information or recovering the object
to notify the launching state and to return the object or hold it at the disposal of
representatives of the launching authority. On jurisdiction over space activities
and spacecraft, see M. McDougal, H. Lasswell, and I. Vlasic, supra note 9, at
646–748.

245 International Rules for the Prevention of Collisions at Sea, the Navigation Rules
for the Great Lakes and their Connecting and Tributary Waters, the Navigation
Rules for the Red River of the North and Rivers Emptying into the Gulf of
Mexico and their Tributaries, and the Navigation Rules for Harbors, Rivers and
Inland Waters Generally. In addition, there are three sets of pilot rules, local

ordinances and regulations, customs, etc. (J. Lucas, *Admiralty* 746–7 (1969)). On collisions, see G. Gilmore and C. Black, *The Law of Admiralty* 395–442 (1957). For the new international regime, see Convention on International Regulations for Preventing Collisions at Sea, *done* 20 October 1972, T.I.A.S. no. 8587.

246 See, eg, Convention for the Unification of Certain Rules with Respect to Assistance and Salvage at Sea, 23 September 1910, 37 Stat. 1658 (1911–13), T.S. no. 576; International Convention for the Safety of Life at Sea, *done* 17 June 1960, [1965] 1 U.S.T. 185, T.I.A.S. no. 5780, 536 U.N.T.S. 27 and the new convention on that subject, supra note 102; Convention on Facilitation of International Maritime Traffic, *done* 9 April 1965, [1967] 1 U.S.T. 410, T.I.A.S. no. 6251, 591 U.N.T.S. 265.

247 Supra note 65, art. 2. See also note 15 supra and Convention for Protection of Submarine Cables, 14 March 1884, 24 Stat. 989 (1885–87), T.S. no. 380.

248 Convention for the Unification of Certain Rules Relating to International Transportation by Air, 12 October 1929, 49 Stat. 3000 (1935–6), T.S. no. 876, 137 L.N.T.S. 11; International Air Services Transit Agreement, supra note 243; Convention on International Civil Aviation, *done* 7 December 1944, 61 Stat. 1180 (1947), T.I.A.S. no. 1591, 15 U.N.T.S. 295 respectively

249 Convention for the Suppression of Unlawful Seizure of Aircraft, *done* 16 December 1970, [1971] 2 U.S.T. 1641, T.I.A.S. no. 7192; Convention for the Suppression of Unlawful Acts against the Safety of Civil Aviation, *done* 23 September 1971, [1971] 1 U.S.T. 565, T.I.A.S. no. 7570

250 Supra notes 16 and 110

251 *Done* 12 November 1965, [1967] 1 U.S.T. 575, T.I.A.S. no. 6267

252 Agreement Relating to the International Satellite Communications Organization, *done* 20 August 1971, [1972] 4 U.S.T. 3813, T.I.A.S. no. 7532. See also Operating Agreement Relating to the International Telecommunications Satellite Organization, *done* 20 August 1971, [1972] 4 U.S.T. 4091, T.I.A.S. no. 7532; Convention on the International Maritime Satellite Organization, *done* 3 September 1976, in 15 *Int'l Legal Materials* 1051 (1976).

253 42 U.S.C.A. § 7401 *et seq.* (West Supp. 1977)

254 For the environmental legislation of major industrial countries and other international environmental legal information, see BNA, *Int'l Environmental Guide* [binder service].

255 Supra note 226

256 *Adopted* 16 November 1972, [1975] 1 U.S.T. 37, T.I.A.S. no. 8226. As was the Endangered Species Convention, the World Heritage Convention was concluded under the auspices of the U.N. Educational, Scientific and Cultural Organization.

257 3 February 1971, in 11 *Int'l Legal Materials* 969 (1972). This convention was concluded at an International Conference on the Conservation of Wetlands and Waterfowl held in and at the urgence of Iran.

258 World Heritage Convention, supra note 256, art. 4

259 Ibid, parts 3, 4, and 5 respectively. The Committee and the Fund are set up within UNESCO.

260 Wetlands Convention, supra note 257, arts. 2–3

261 Ibid, art. 6. Provision is made for the International Union for the Conservation of Nature and Natural Resources to perform the continuing bureaucratic duties under this convention.

262 *Report*, supra note 3, at 5

263 Supra note 131

264 Ibid, art. 8, enumerates the following order of precedence of uses: 1 / uses for domestic and sanitary purposes; 2 / uses for navigation, including the service of canals for the purposes of navigation; 3 / uses for power and for irrigation. On the IJC and its jurisdiction, see ibid, arts. 7–12.

265 Supra note 135

266 Ibid, arts. 2–10

267 On the Rhine and Danube and their commissions, see Stein, The Potential of Regional Organizations in Managing Human Environment 23–6, 49–53 (Woodrow Wilson Int'l Center for Scholars Env. Ser. 202, 1972) (same article as supra note 146, with annexes – see also that note).

268 See 59–60 supra.

269 See *Report*, supra note 3, at 59.

270 On the initiation of such activities see generally, The Environment Programme: Approval of Activities within the Environment Programme, in the Light, inter alia, of Their Implications for the Fund Programme, U.N. Doc. UNEP/GC/14/Add.2 (1973); Review of the Environmental Situation and of Activities Relating to the Environment Programme, U.N. Doc. UNEP/GC/30 (1975).

271 G.A. Res. 3054, 28 U.N. GAOR Supp. 30, at 38, U.N. Doc. 9030 (1973)

272 See Consideration of the Economic and Social Situation in the Sudano-Sahelian Region Stricken by Drought and Measures to Be Taken for the Benefit of that Region: Report of the Secretary-General, U.N. Doc. E/5457 (1974).

273 Convention Establishing a Permanent Inter-State Drought Control Committee for the Sahel, *done* 12 September 1973, in 13 *Int'l Legal Materials* 537 (1974). The parties are Chad, Mali, Mauritania, Niger, Senegal, and Upper Volta.

274 See *Report of the United Nations Conference on Desertification*, U.N. Doc. A/Conf.74/36 (1977). The Governing Council of the U.N. Environment Programme served as the intergovernmental preparatory body for this conference. For an early indication of the priority of these matters within UNEP, see

Report of the UNEP GC (2d session), supra note 217, at 90–1. For the particular efforts in regard to the Sahel, see Measures to be Taken for the Benefit of the Sudano-Sahelian Region, U.N. Doc. UNEP/GC.6/9/Add.2 (1978).

275 *Report of Habitat: United Nations Conference on Human Settlements*, U.N. Doc. A/Conf.70/15, at 2–3 (1976). This declaration recalled the prior work of several U.N. bodies, including 'the decisions of the United Nations Conference on the Human Environment, as well as the recommendations of the World Population Conference, the United Nations World Food Conference, the Second General Conference of the United Nations Industrial Development Organization, the World Conference of the International Women's Year; the Declaration and Programme of Action adopted by the sixth special session of the General Assembly of the United Nations and the Charter of Economic Rights and Duties of States that establish the basis of the New Economic Order' (ibid, at 2). Habitat also passed a resolution on the United Nations Water Conference (ibid, at 115).

Simultaneously with the United Nations Conference, a non-governmental Habitat Forum was held at Jerico Beach, Vancouver. More than five thousand participants from ninety countries took part. See ibid, at 182–3.

276 The recommendations for national action and for international co-operation are found in ibid, at 10–104. On institutional arrangements for international co-operation in the field of human settlements, see G.A. Res. 32/162 (19 December 1977).

277 *Report*, supra note 3, at 2

278 See 71 infra.

279 G.A. Res. 3201, U.N. GAOR 6th Special Sess. Supp. 1, at 3 and 4, U.N. Doc. A/9559 (1974)

280 G.A. Res. 3281, 29 U.N. GAOR Supp. 31, at 50, U.N. Doc. A/9631 (1974). Art. 30 deals specifically with state responsibility for the environment.

281 Declaration of the U.N. Conference on the Human Environment, *Report*, supra note 3, at 2

282 Ibid, at 3

283 Ibid, at 2. See also *Development and Environment* (Report and Papers of a Panel of Experts Convened by the Secretary-General of the UNCHE – the 'Founex Report,' 1971); Development and Environment, U.N. Doc. A/Conf.48/10 (1971).

284 Recommendations 102–9, *Report*, supra note 3, at 54–8

285 Both a UNEP working paper and the report of the expert group are annexed to Environment and Development Including Irrational and Wasteful Use of Natural Resources and Ecodevelopment, U.N. Doc. UNEP/GC/102 (1977).

235 Notes to chapter 3, pages 69–71

286 On the creation of the UN Environment Fund, see chapter 2, 8. For the most
recent annual report, see Report on the Implementation of the Fund Programme
in 1977, U.N. Doc. UNEP/GC.6/13 (1978).

On the Habitat and Human Settlements Foundation, see 65–6 supra. The idea
of a human settlements foundation or fund of some sort has been developing for
quite some time. It was first proposed in Stockholm Recommendation 17, *Report*, supra note 3, at 14. The proposal has received further study by the UNEP
secretariat and won the support of the Governing Council of UNEP. See Establishment of an International Fund or Financial Institution for Human Settlements (G.A. Res. 2999 [XXVII]) – Report of the Secretary-General, UN Doc.
UNEP/GC/19 (1974); *Report of the UNEP GC (2d session)*, supra note 217, at
44–9, 113–17. For the actual establishment of the UN Habitat and Human
Settlements Foundation, see G.A. Res 3329, 29 U.N. GAOR Supp. 31, at 64, U.N.
Doc. A/9631 (1974). See also Programme of Operations for the United
Nations Habitat and Human Settlements Foundation, U.N. Doc. UNEP/GC/35
(1975).

287 In ICNT art. 21(2), in 8 *Third UNCLOS: Official Records*, supra note 2, at 8, it
is provided that coastal state laws and regulations relating to innocent passage shall not apply to the design, construction, manning, or equipment of
foreign ships 'unless they are giving effect to generally accepted international
rules or standards.' But see ICNT art. 212(3), ibid, at 37, which may be less
restrictive. The provisions on dumping and pollution from sea-bed activities
are found in arts. 211 and 209 respectively, ibid, at 36, and the double standard concerning pollution from land-based sources is contained in art. 208(4),
ibid.

288 For discussion, see Sachs, Environmental Quality Management and Development Planning: Some Suggestions for Action, in *Development and Environment*, supra note 283, at 123.

289 Principle 12, *Report*, supra note 3, at 5, reads: 'Resources should be made
available to preserve and improve the environment, taking into account the
circumstances and particular requirements of developing countries and any
costs which may emanate from their incorporating environmental safeguards
into their development planning and the need for making available to them,
upon their request, additional international technical and financial assistance
for this purpose.'

290 See ICNT arts. 69–72, in 8 *Third UNCLOS: Official Records*, supra note 2, at
15–16.

291 Declaration on the Human Environment, preamble, *Report*, supra note 3, at 2

292 Ibid, at 4

293 On human rights and nationality, see McDougal, Lasswell, and Chen, Nationality and Human Rights: The Protection of the Individual in External Arenas, 83 *Yale L.J.* 900 (1974).
294 G.A. Res. 217A, at 71, art. 15, U.N. Doc. A/777 (1948)
295 Ibid
296 Annex to G.A. Res. 2200A, 21 U.N. GAOR Supp. 16, at 49, U.N. Doc. A/6316 (1966), reprinted in 61 *Am. J. Int'l L.* 861 (1967)
297 4 November 1950, Eur. T.S. no. 5, 213 U.N.T.S. 221
298 A list of human rights conventions and references is found in 13 M. Whiteman, supra note 7, at 672–4.
299 *Report of the World Population Conference*, U.N. Doc. E/Conf.60/19, at 17 (1974)
300 Besides the demographic projections, the legal and political literature on population is increasing greatly. See, eg, Claxton, The Development of Institutions to Meet the World Population Crisis, 65 *Dept. State Bull.* 165 (1971); P. Ehrlich and A. Ehrlich, *Population, Resources, Environment* (2d ed 1972); Nanda, The Role of International Law and Institutions Toward Developing a Global Plan of Action on Population, 3 *Denver J. Int'l L. & Policy* 1 (1973); *Population: A Clash of Prophets* (E. Pohlman ed 1973); *The World Population Crisis: Policy Implications and the Role of Law* (Proc. Am. Soc. Int'l L. and John Basset Moore Soc. Int'l L. Symposium 1971); Contributions to Population Policy, 26 *Int'l Soc. Sci. J.* No. 2 (1974).
301 Principle 16, *Report*, supra note 3, at 6
302 For the genesis of this UN conference on population, see ECOSOC Res. 1484, 48 U.N. ECOSOC Supp. 1, at 2, U.N. Doc. E/4832 (1970). On the denomination of 1974 as World Population Year, see ECOSOC Res. 1485, ibid, at 3; G.A. Res. 2683, 25 U.N. GAOR Supp. 28, at 55, U.N. Doc. A/8028 (1970). See also ECOSOC Res. 1672A–D, 52 U.N. ECOSOC Supp. 1, at 6, U.N. Doc. E/5183 (1972). For the conference report, see note 229 supra.
303 *Report of the World Population Conference*, supra note 299, at 10
304 Ibid, at 7
305 Ibid, at 25
306 Ibid, at 49
307 Ibid, at 4–51

CHAPTER FOUR

1 *Report of the United Nations Conference on the Human Environment* U.N. Doc. A/Conf.48/14, at 59 (1972) (hereinafter cited as *Report*). See also An Action Plan for the Human Environment, U.N. Doc. A/Conf.48/5 (1972); In-

ternational Council of Scientific Unions, Scientific Committee on Problems of the Environment, Global Environmental Monitoring (1971).

2 *Report of the Governing Council of the United Nations Environment Programme on Its First Session*, 28 U.N. GAOR Supp. 25, at 6, 45–6, U.N. Doc. A/9025 (1973). See also Action Plan for the Human Environment: Programme Development and Priorities – Report of the Executive Director, U.N. Doc. UNEP/GC/5 (1973).

3 For the basic blueprint, see Report of the Intergovernmental Meeting on Monitoring, U.N. Doc. UNEP/GC/24 (1974). For the time goal, see *Report of the Governing Council of the United Nations Environment Programme on the Work of Its Fifth Session* 32 U.N. GAOR Supp. 25, at 9, U.N. Doc. A/32/25 (1977). See also The Global Environmental Monitoring System, U.N. Doc. UNEP/GC/31/Add.2 (1975); Development and Implementation of the Global Environmental Monitoring System, U.N. Doc. UNEP/GC/Inf.2 (1977).

4 Informal Composite Negotiating Text (ICNT) arts. 205–6, in 8 *Third United Nations Conference on the Law of the Sea: Official Records* 36 (1977). An earlier text required that such reports be communicated 'to UNEP'; subsequently, however, the Conference decided generally to avoid referring to organizations by name in the draft articles. See Results of Consideration of Proposals and Amendments Relating to the Preservation of the Marine Environment, in ibid, vol. 4, at 200 (1975). See also Global Environmental Monitoring System of the United Nations Environment Programme, in ibid, at 207.

5 For delineation of UNEP's priority areas, see *Report of the UNEP GC (1st session)*, supra note 2, at 39–43. On UNEP's overall aims and operations, see The Environment Programme (Levels One, Two and Three), U.N. Doc. UNEP/GC/90 and Add. 1 and 2 (1977).

As regards the particular case of the Mediterranean, the framework is provided by the Convention for the Protection of the Mediterranean Sea against Pollution, *done* 16 February 1976, in 15 *Int'l Legal Materials* 290 (1970). The original Action Plan is reprinted in ibid, vol. 14, at 467 (1975). For discussion of this and other UNEP regional efforts, see chapter 3, 38.

6 42 U.S.C. § 4321, 4332(2) (C) (1970). In preparing the environmental impact statements required by this section, federal officials are required to 'recognize the worldwide and long-range character of environmental problems' and promote international co-operation in 'anticipating and preventing a decline in the quality of mankind's world environment' (ibid, § 4332(E)).

7 The Environmental Programme: Approval of Activities within the Environment Programme, in the Light, inter alia, of Their Implications for the Fund Programme, U.N. Doc. UNEP/GC/14/Add.2, at 103 (1973)

8 Ibid, at 105–7

9 More specifically, the Executive Director has proposed among the goals for UNEP by 1982 '[a]n operational International Referral System (IRS) with nearly all countries having registered sources and making use of the service' and '[t]he International Register of Potentially Toxic Chemicals (IRPTC) in a position to issue warnings and technical publications.' *Report of the UNEP GC (5th session)*, supra note 3, at 9. See also Progress Report on International Referral System Development, U.N. Doc. UNEP/GC/ Inf.7 (1977). On the Register, see IRPTC *Bulletin*.

10 *Done* 23 May 1972, in 11 *Int'l Legal Materials* 761, art. 3 (1972)

11 19 June 1973, in ibid, vol. 12, at 911, art. 2(a) (1973)

12 *Done* 9 May 1974, in 13 *Int'l Legal Materials* 598 (1974)

13 See chapter 3, 51.

14 G.A. Res. 2995, 27 U.N. GAOR Supp. 30, at 42, U.N. Doc. A/8730 (1972); G.A. Res. 3129, 28 U.N. GAOR Supp. 30, at 48, U.N. Doc. A/9030 (1973)

15 See Co-operation in the Field of the Environment Concerning Natural Resources Shared by Two or More States, U.N. Doc. UNEP/GC/44 (1975).

16 Report of the Intergovernmental Working Group of Experts on Natural Resources Shared by Two or More States on the Work of Its Fifth Session, U.N. Doc. UNEP/IG.12/2, at 9 (1978)

17 Beesley, Editorial Comment: The Missing Environmental Perspective, 1 *Earth L.J.* 87 (1975). He added, in line with the overall thesis of the present study, the observation that 'this switch in attention seems to ignore the symbiotic relationship between resource depletion, energy consumption, and the quality of the environment' (ibid, at 88). See also Beesley, New Legal Environment, in *Future of the Offshore: Legal Developments and Canadian Business* 3,5 (D. Patton, C. Beckton, and D. Johnston eds. 1977).

18 *Report*, supra note 1, at 6

19 Ibid, at 60

20 For discussion, see Revolving Fund (Information), U.N. Doc. UNEP/GC/47 (1975).

21 For the most recent, see State of the Environment Report, U.N. Doc. UNEP/GC.6/4 (1978). In accordance with a directive of the Governing Council, it is intended that the annual state of the environment report should be selective and focus on specific subjects, and that there should be an analytic, comprehensive assessment submitted each fifth year.

22 See the Swedish letter to the United Nations Secretary-General, U.N. Doc. E/4466/Add.1 (1968). The subsequent invitation of Sweden to hold the Conference in Stockholm is contained in U.N. Doc. A/7514 (1969).

23 See chapter 3, 27, 35. See also 85–6 infra.

24 See chapter 3, 35–6 on the two conventions and 53 on the environmental assessment draft article.

25 For evidence of these activities with particular regard to the marine environment, see the Kenyan draft articles on the protection of the marine environment, U.N. Docs. A/AC.138/SC.III/L.41 (1973) and A/Conf.62/C.3/L.2 (1974), the latter reprinted in 3 *Third United Nations Conference on the Law of the Sea: Official Records* 245 (1975).

26 Their absence was in protest against the use of the 'Vienna formula' of conference representation, which at the time allowed West Germany but not East Germany to attend.

27 Environment Stockholm 17 (U.N. CESI 1972)

28 UNEP, Rules of Procedure of the Governing Council, U.N. Doc. UNEP/GC/3/Rev.1, at 32, Rule 69 (1974). On the role of such organizations, see Kay and Skolnikoff, International Institutions and the Environment Crisis, 26 *Int'l Org.* no. 2 (1972).

29 D. Meadows, D. Meadows, J. Randers, and W. Behrens, *The Limits to Growth* (1972); M. Mesarovic and E. Pestel, *Mankind at the Turning Point* (1974)

30 See, eg, the papers by the University of Sussex Science Policy Research Unit, The Limits to Growth, 5 *Futures* nos. 1–2 (1973); The No-Growth Society, 102 *Daedulus* (Fall 1973).

31 See SCEP, Man's Impact on the Global Environment (1970); SCOPE, supra note 1; SMIC, Inadvertant Climate Modification (1971). ICSU also has other committees working on environmental problems.

32 IIEA, World Energy, the Environment and Political Action (based on a workshop co-sponsored with the Aspen Institute for Humanistic Studies 1971)

33 Hallman, Towards an Environmentally Sound Law of the Sea (IIED 1974)

34 B. Ward and R. Dubos, *Only One Earth* (1972); H. Sprout and M. Sprout, *Toward a Politics of the Planet Earth* (1971); R. Falk, *This Endangered Planet* (1971). It should also be added that Barbara Ward has gained further note as an environmentalist for her leadership of the International Institute for Environment and Development.

35 For discussion and citations, see chapter 3, 32–3 and n59; 33–4 and nn63, 64, and 71; 34–5 and n72; 35, 36 and n76; 35–6 and nn79 and 88; 36–7 and n92; and 37–8 and n96 respectively. For a summary of states which have accepted several of these conventions and their date of entry into force (if any), see IMCO, Status of International Conventions relating to Marine Pollution of which IMCO Is Depository or Is Responsible for Secretariat Duties, U.N. Doc. MEPC IX/2 (18 March 1978). See generally Register of International Conventions and Protocols in the Field of the Environment, U.N. Doc. UNEP/GC/Inf.5 (1977) and Supp. (1978); International Conventions and Protocols in the Field of the Environment, U.N. Doc. UNEP/GC.6/8 (1978). But see Contini and Sand, Methods to Expedite Environment Protection: International Ecostandards, 66 *Am. J. Int'l L.* 37 (1972).

36 See chapter 3, 40 and n115; 39 and n7; and 39–40 and n112 respectively.

37 See chapter 3, 62, 58 and nn255 and 226; 58 and n219; and 58 and n224.
38 *Report*, supra note 1, at 2, 59, 61, 66–8
39 See *Report of the World Population Conference*, U.N. Doc. E/Conf.60/19 (1974).
40 G.A. Res. 2994–3004, 27 U.N. GAOR Supp. 30, at 42–8, U.N. Doc. A/8730 (1972). See also G.A. Res. 3123, 3128–33, 28 U.N. GAOR Supp. 30, at 46–51 (1973); G.A. Res. 3225–7, 29 U.N. GAOR Supp. 31, at 60–2, U.N. Doc. A/9631 (1974).
41 See *Reports of the UNEP GC*, supra notes 2 and 3 and annually, Annex 1 of each.
42 For discussion see Bilder, The Role of Unilateral State Action in Preventing International Environmental Injury (U. Wisc. Sea Grant College Program WIS–SG–73–219, 1973). See generally 1 Oppenheim, *International Law* 25 (8th ed Lauterpacht 1955).
43 R.S.C. 1970 (1st Supp.) c. 2 and 45 (1970), reprinted in 9 *Int'l Legal Materials* 543 and 553 (1970). Contemporaneously with the passage of these acts, Canada modified its declaration under article 36 of the Statute of the International Court of Justice to decline compulsory jurisdiction as regards issues arising out of its antipollution measures. See Canadian Declaration Concerning the Compulsory Jurisdiction of the ICJ, ibid, at 598.
44 See Beesley, Rights and Responsibilities of Arctic Coastal States: The Canadian View, 3 *J. Maritime L. & Com.* 1 (1971) and his The Arctic Pollution Prevention Act: Canada's Perspective, 1 *Syr. J. Int'l L. & Com.* 226 (1973); Legault, Canadian Arctic Waters Pollution Prevention Legislation, in *The Law of the Sea: The United Nations and Ocean Management* 294 (Proc. 5th Annual Conf. Law of the Sea Institute 1970) and his Maritime Claims, in *Canadian Perspectives on International Law and Organization* 377 (R. Macdonald, G. Morris, and D. Johnston eds 1974). But see Bilder, The Canadian Arctic Waters Pollution Prevention Act: New Stresses on the Law of the Sea, 69 *Mich. L. Rev.* 1 (1970); Henkin, Arctic Anti-Pollution: Does Canada Make – or Break – International Law? 65 *Am. J. Int'l L.* 131 (1971).
45 U.S. Statement on Canada's Proposed Legislation, 9 *Int'l Legal Materials* 605, 605–6 (1970)
46 Canadian Reply to the U.S. Government, ibid, at 607, 607–10
47 Canadian Order in Council, [1971] P.C. no. 366. The term *fisheries closing lines* does not appear either in the Territorial Sea and Fishing Zones Act which authorizes the action, supra note 43, or in the Order in Council; it is, however, in common use.
48 U.S. Regrets Canada's Extension of High Seas Jurisdiction, 64 *Dept. State Bull.* 139 (1971)
49 Canada's Action on Fisheries Closing Lines Opposed by U.S., ibid, at 448
50 R.S.C. 1970 (2d Supp.) c. 27 (1971). The amendments are now Part 20 of the Canada Shipping Act. For discussion, see 88 infra.

51 33 U.S.C.A. § 1221 *et seq.* (Supp. V 1975). The purpose of this act is stated as the
prevention of damage to various structures and 'to protect the navigable
waters and the resources therein from environmental harm ... ' (ibid § 1221).
The standard-setting competence of the Secretary is defined in §§1221–2. See
94 infra. See also the U.S. Deepwater Ports Act of 1974, 33 U.S.C.A. § 1501 *et seq.*
(West Supp. 1978), whose purposes, inter alia, are to 'authorize and regulate
the location, ownership, construction and operation of deepwater ports in
waters beyond the territorial limits of the United States,' and in so doing to
'provide for the protection of the marine and coastal environment' (ibid, §
1501). On this latter question, see Knight, International Legal Aspects of Deep
Draft Harbor Facilities, 4 *J. Maritime L. & Com.* 367 (1973).

52 S. 2801 and H.R. 13904, 92d Cong., 2d Sess. (1972); reintroduced as S. 1134 and H.R.
9, 93d Cong., 1st Sess. (1973); see also Amendments to the latter, 93d Cong., 2d
Sess. (1974). Senator Metcalf was Chairman of the Subcommittee on Minerals,
Materials and Fuels of the Senate Committee on Interior and Insular Affairs.

   The latest version of the bill introduced by the late Senator is S. 2053, 95th
Cong., 1st Sess. (1977). The House of Representatives most recently considered
H.R. 12988, 95th Cong., 2d Sess. (1978), which is the substitute for the original H.R.
3350, 95th Cong., 1st Sess. (1977). Federal investment guarantees are a central
issue that stalled passage of the legislation. The proponents say that the minerals
are needed now and the guarantees are essential for the mining companies to
obtain financing and proceed to the next stage of development of sea-bed mining;
their opponents, on the other hand, claim that the guarantees would be a 'bail out'
or 'blank cheque' for the companies and question the very concept of indemnifying
US miners against losses incurred as a consequence of a treaty both the President
and the Senate believe to be in the overall best interest of the United States. For
discussion, see [1978] *Cong. Q.* 124–6.

53 For some commentary on Third World reactions, see Alexander, Dead Ahead
Toward a Bounded Main, *Fortune* 129, 208 (October 1974). Compare Oelsner,
US Eases Stand on Seabed Mining, *N.Y. Times*, 12 August 1972, at 5, col. 1.

54 The Deepsea Ventures 'Notice of Discovery ... ' appears in 14 *Int'l Legal Materials* 51 (1975). For some responses refusing to recognize such exclusive mining
rights by the US Department of State and the Canadian, Australian and British
governments, see ibid, at 66, 67, 795, 796.

   On the *Glomar Explorer*, see Hersh, CIA Salvage Ship Brought up Part of
Soviet Sub Lost in 1968, Failed to Raise Atom Missiles, *N.Y. Times*, 19 March
1975, at 1, col. 8. For some US and international reactions, see Hersh, 3 Panels
in Congress Plan Inquiries into Sub Salvage, ibid, 21 March 1975, at 1, col. 1;
Submarine Project Affects Big Powers at Sea-Law Meeting, ibid, 21 March
1975, at 30, col. 4. See also Alpern et al., CIA's Mission Impossible, 85 *News-*

week 24 (31 March 1975); Did Hughes Really Build a Mining Ship? *Bus. Week* 26 (7 April 1975). Compare with the earlier Now Howard Hughes Mines the Ocean Floor, *Bus. Week* 47 (16 June 1973) and the note on this article in People and Places, *N.Y. Times*, 15 June 1973, at 51, col. 3.

55 The international 'Area' is defined as 'the sea-bed and ocean floor and subsoil thereof beyond the limits of national jurisdiction' (ICNT art. 1(1), in 8 *Third UN-CLOS: Official Records*, supra note 4, at 6). The regime governing the area is established by arts. 133–92, ibid, at 22–34. See also arts. 86–120 on the high seas, ibid, at 17–21.

The diplomatic mood at UNCLOS III is at present one of cautious optimism, but the institutional arrangements are still far from final. What is emerging is neither the parallel system of exploitation that some states want nor an alternative unified system. The ICNT blueprint has highly detailed specifications not only of international machinery, but of production controls for land-based producers of nickel, transfer of sea-bed exploration and exploitation technology, and periodic and comprehensive review of the system. At the Geneva portion of the seventh session of UNCLOS III, amendments were agreed to for arts. 150 bis, Annex II para. 4(c) (ii), and art. 153 of the ICNT dealing with the last-mentioned subjects. See Revised Suggested Compromise Formula by the Chairman of Negotiating Group 1, U.N. Doc. NG1/10/Rev.1 (16 May 1978), reprinted in *Reports of the Committees and Negotiating Groups on Negotiations at the Seventh Session Contained in a Single Document Both for the Purposes of Record and for the Convenience of Delegations* 4, 5, 9–10 (19 May 1978).

56 Interim Fisheries Zone Extension and Management Act of 1973, S. 1988, 93d Cong., 1st Sess. (1973). Senator Magnuson introduced the bill for himself and Senators Cotton, Hollings, Jackson, Pastore, and Stevens. See also S. 380 and S. 2338 introduced during the same session.

57 120 *Cong. Rec.* 29492, 29493 (1974). Compare Moore, Statement on Proposed U.S. Legislation Extending Fisheries Jurisdiction, 13 *Int'l Legal Materials* 1292 (1974).

58 16 U.S.C.A. § 1801 *et seq.* (West Supp.1978). The prohibition of foreign fishing in the two-hundred-mile fishery conservation zone came into effect on 1 March 1977. For discussion of the acceptance of two-hundred-mile exclusive economic zones by UNCLOS III, see chapter 3, 24–5.

59 S. 1341, 94th Cong., 1st Sess. 1974. The Federal Water Polllution Control Act is found at 33 U.S.C. § 466 *et seq.* (1970),and the Ports and Waterways Safety Act is cited at note 51 supra.

60 Supra notes 43 and 50 respectively

61 121 *Cong. Rec.* 8694, 8694–5 (1975)

62 Pub. L. no. 95–217, 91 Stat. 1566 (1977). Section 58 extended various provisions of § 311(b) of the Federal Water Pollution Control Act, supra note 59, § 1362(7), dealing with oil and hazardous substances liability.

63 *U.K. Stat.* c. 21 (1971). The Oil in Navigable Waters Act defines offences and establishes liability for oil pollution under various circumstances. The radical nature of the legislation is the provision that the British government may by Order in Council, and 'in such cases and circumstances as may be specified in the Order,' apply the act to a ship a / which is not a ship registered in the United Kingdom, and b / which is for the time being outside the territorial waters of the United Kingdom (ibid, § 8(10)). This has been done in the Oil in Navigable Waters (Shipping Casualties) Order 1971, reprinted in 1 *New Directions in the Law of the Sea* 229 (S. Lay, R. Churchill, and M. Nordquist eds 1973). See also The Prevention of Oil Pollution Act 1971, c. 60; The Merchant Shipping (Oil Pollution) Act 1971, c. 59. For a brief discussion of this British legislation, see Johnston, Marine Pollution Control: Law, Science and Politics, 28 *Int'l J.* 69, 85 (1972).

64 See Hayashi, Comparative National Legislation on Offshore Pollution, 1 *Syr. J. Int'l L. & Com.* 250, 253–6 (1973). In addition to the legislation here, Hayashi discusses the Japanese Marine Pollution Prevention Law of 1970 and the US Water Quality Improvement Act of 1970 in terms of their offshore reach.

65 The Canada Shipping Act, supra 86, 88 and note 50, as amended by Part 20, establishing comprehensive pollution jurisdiction, applies to 'all Canadian waters' and to 'any fishing zones of Canada' (ibid, § 727). Consequently, the recent extension of Canada's fishing zones to two hundred miles can be argued to have had the effect of extending antipollution jurisdiction out to that distance, when read in conjunction with Part 20. See Fishing Zones of Canada (Zones 4 and 5) Order, 111 Can. Gaz., Part 2, at 115 (1977); Fishing Zones of Canada (Zone 6) Order, ibid, at 652. Jurisdiction over dumping of wastes and other substances could be said similarly to have been indirectly extended, since Canada's Ocean Dumping Control Act also applies to 'any fishing zones' (1 Can. Stat. (1974–75–76), c. 55 (1975)).

66 See also Regulations Respecting the Prevention of Pollution of Water by Pollutants from Ships, 107 Can. Gaz., Part 2, at 414 (1973); Oil Pollution Prevention Regulations, ibid, vol. 105, Part 2, at 1723 (1971), and ibid, vol. 107, Part 2, at 2238 (1973).

67 For discussion of the present state of the negotiations on protection and preservation of the marine environment at UNCLOS III, see chapter 3, 25–7.

68 Article 34 of the ICJ Statute states the limitation: 'Only States may be parties before the Court.' In its opinion on *Reparations for Injuries Suffered in the Service of the United Nations*, [1949] I.C.J. 174, the Court held that this compe-

tence extends to international organizations created by states. The rules of the Permanent Court of International Arbitration have, however, been amended to provide for non-state parties (Rules of Arbitration and Conciliation for Settlement of International Disputes Between Two Parties Only One of Which Is a State, in 57 *Am. J. Int'l L.* 500 (1963)).

69 (*Liechtenstein* v *Guatemala*), [1955] I.C.J. 4

70 25 I.L.R. 91 (Italian-United States Conciliation Commission 1963), digested in 53 *Am. J. Int'l L.* 944 (1963)

71 *Barcelona Traction, Light and Power Company, Ltd.* (*Belgium* v *Spain*), [1970] I.C.J. 3

72 Ibid, at 49

73 *Done* 29 March 1972, [1973] 2 U.S.T. 2389, T.I.A.S. no. 7762, art. 8. For discussion, see chapter 3, 39–40.

74 Convention, supra note 73, art. 14

75 *Done* 29 November 1969, in 9 *Int'l Legal Materials* 45, art. 9 (1969). See chapter 3, 33–4.

76 Convention on the Protection of the Environment, *done* 19 February 1974, in ibid, vol. 13, at 591, art. 3 (1974). See chapter 3, 53.

77 *Wilderness Soc'y* v *Morton*, 463 F.2d 1261 (D.C. Cir. 1972)

78 Application Instituting Proceedings on Behalf of Australia, quoted in ICJ Interim Protection, Order of 22 June 1973, [1973] I.C.J. 99, 103. For discussion of the cases see chapter 3, 41–2.

79 Application of New Zealand, quoted in the other Interim Protection, Order of 22 June 1973, [1973] I.C.J. 135, 139

80 Supra note 68. Having in mind the death of Count Bernadotte while serving as a United Nations Mediator in Palestine, the UN General Assembly requested an advisory opinion from the ICJ: 'In the event of an agent of the United Nations in the performance of his duties suffering injury in circumstances involving the responsibility of a State, has the United Nations, as an Organization, the capacity to bring an international claim against the responsible *de jure* or *de facto* government with a view to obtaining the reparation due in respect of the damage caused a / to the United Nations, b / to the victim or to persons entitled through him?' ([1949] I.C.J. 174, 175). The ICJ gave an affirmative answer to both parts. In response to the further question 'How is action by the United Nations to be reconciled with such rights as may be possessed by the State of which the victim is a national?' (ibid), the Court explained that the action is based not upon the nationality of the victim but upon his status as an agent of the organization.

81 For my further opinions on this point, see Note, New Perspectives on International Environmental Law, 82 *Yale L. J.* 1659 (1973), reprinted in 5 *Environ-*

ment L. Rev. 669 (1974). See also McDougal and Schneider, The Protection of the Environment and World Public Order: Some Recent Developments, 45 *Miss. L. J.* 1085 (1974).

82 *Done* 12 May 1954, [1961] 3 U.S.T. 2989, T.I.A.S. no. 4900, 327 U.N.T.S. 3; with amendments *adopted* 11 April 1962, [1966] 2 U.S.T. 1523, T.I.A.S. no. 6109, 600 U.N.T.S. 332; 21 October 1969, T.I.A.S. no. 8505; 15 October 1971, in 11 *Int'l Legal Materials* 267 (1972)

83 Goldie, Development of an International Environmental Law – An Appraisal, in *Law, Institutions and the Global Environment* 104, 118 (J. Hargrove ed 1972)

84 These are the prosecution of the owners of the *Texaco Mississippi* by the United States Coast Guard in October 1966 for discharging oil contrary to the convention and the prosecution of the owners and master of *The Huntington* by the United Kingdom in October 1972 for their violations. In the latter case, the owner was fined £2500 plus £1000 costs, and the master was fined £250. Both prosecutions were the result of citing reports by crews of Canadian Forces aircraft. See Beesley, The Canadian Approach – Environmental Law on the International Plane, in *Private Investors Abroad – Problems and Solutions in International Business in 1973*, at 239, 247 n18 (Proc. Southwestern Legal Fnda. 1973).

85 Draft Environmental Impact Statement on the Law of the Sea, prepared by the United States, May 1974, cited in Hallman, supra note 33, at 57 n1

86 International Convention Relating to Intervention on the High Seas in Cases of Oil Pollution Casualties and International Convention on Civil Liability for Oil Pollution Damage, *done* 29 November 1969, [1975] 1 U.S.T. 765, T.I.A.S. no. 8068; International Convention on the Establishment of an International Fund for Compensation for Oil Pollution Damage, *done* 18 December 1971, in 11 *Int'l Legal Materials* 284 (1972). For discussion see chapter 3, 33–4.

87 *Report*, supra note 1, Annex III, at 3, principle 20. See chapter 3, 35.

88 See generally B. Boczek, *Flags of Convenience* (1962). Ships are registered in Panama, Liberia, or Honduras ('Panlibhon' ships) and occasionally elsewhere to avoid wage or tax legislation in the countries of their owners. Even were such advantages to disappear, the continuance or expansion of the practice to avoid enforcement of antipollution or other environmental legislation is not unimaginable. See also OECD Study on Flags of Convenience, 4 *J. Maritime L. & Com.* 231 (1973).

89 Multiple jurisdictions are not unknown to international law. The traditional law of the sea, for example, provides for universal jurisdiction for punishment of crimes of piracy. See M. McDougal and W. Burke, *The Public Order of the Oceans* 875–9 (1962). Concurrent jurisdiction of this kind is, however, something new for international environmental law.

90 Supra note 51
91 33 U.S.C. § 1226. See also Part 20 of the Canada Shipping Act, supra note 50.
92 33 U.S.C. § 1226
93 Ibid, § 1226–7
94 Merchant Shipping (Minimum Standards) Convention, *adopted* 29 October 1976, in 15 *Int'l Legal Materials* 1288 (1976). A port state which receives a complaint may prepare a report addressed to the flag state of the vessel concerned, with a copy to the Director General of the ILO, and may also 'take measures necessary to rectify any conditions on board which are clearly hazardous to safety or health' (ibid, art. 4). See also Recommendation concerning the Improvement of Standards in Merchant Ships, *adopted* 29 October 1976 (ibid, at 1293). On acceptance of universal port state enforcement at UNCLOS III, see chapter 3, 27.
95 Supra note 43, § 14 *et seq.*
96 Convention on the Dumping of Wastes at Sea, *done* 13 November 1972, in 11 *Int'l Legal Materials* 1294 (1972); International Convention for the Prevention of Pollution from Ships, *done* 2 November 1973, in ibid, vol. 12, at 1319 (1973). See chapter 3, 36–7.
97 See ibid, arts. 8 and 9 respectively.
98 *Done* 29 April 1958, [1966] 1 U.S.T. 138, T.I.A.S. no. 5969, 559 U.N.T.S. 285, art. 6. See chapter 3, 73.
99 *Report*, supra note 1, at 48; originally contained in Report of the Second Session of the Intergovernmental Working Group on Marine Pollution, U.N. Doc. A/Conf.48/IWGMP.II/5, at 7–8 (1971)
100 Convention on the Prevention of Marine Pollution by Dumping of Wastes and Other Matter, *done* 29 December 1972, [1975] 2 U.S.T. 2403, T.I.A.S. no. 8165, art. 7; Convention for the Prevention of Pollution from Ships, *done* 2 November 1973, in 12 *Int'l Legal Materials* 1319, art. 3 (1973). See chapter 3, 36–7. See also 83 supra.
101 Convention, supra note 100, arts. 7 and 3 respectively
102 ICNT art. 237, in 8 *Third UNCLOS: Official Records*, supra note 4, at 41
103 See chapter 3, 69.
104 See the draft article on technical assistance agreed to by the Third Committee at the Caracas Session of UNCLOS III, now ICNT art. 203, in 8 *Third UNCLOS: Official Records*, supra note 4, at 35. See also chapter 3, 69.
105 Report of the Committee on the Uses of Waters of International Rivers, in International Law Association, *Report of the Fifty-Second Conference Held at Helsinki* 478, 486–8, arts. 4 and 5 (1967). See chapter 3, 44–5.
106 *Report*, supra note 105, at 492, art. 7
107 Ibid, at 493, art. 8

108 Ibid, at 494
109 Draft European Convention for the Protection of International Watercourses Against Pollution in Ad Hoc Committee of Experts to Prepare a European Convention on the Protection of International Fresh Waters Against Pollution, Report, Council of Europe Doc. EXP/Eau (74) 6 Add.1 (13 March 1974). See chapter 3, 45.
110 Art. 150 bis, supra note 55. Canada is a major land producer of nickel and the United States a major importer, so their joint proposal represented a true compromise.
111 See U.N. Declaration on the Human Environment, preamble, *Report*, supra note 1, at 2, quoted in chapter 3, 68.
112 See Principle 12, ibid, at 5, quoted at 235 n289. See chapter 3, 70.
113 G.A. Res. 2997, 27 U.N. GAOR Supp. 30, at 43, 44, U.N. Doc. A/8730 (1972) (emphasis added)
114 *Fisheries Jurisdiction (United Kingdom of Great Britain and Northern Ireland v Iceland)* and *Fisheries Jurisdiction (Federal Republic of Germany v Iceland)*, [1974] I.C.J. 3 and 175. For discussion, see chapter 3, 58–9.
115 Fisheries Jurisdiction (*U.K.* v *Iceland*), supra note 114, at 29.
116 Ibid, at 30
117 Ibid, at 31
118 For some discussion relating to this point, see the separate concurring opinion of Judge Dillard, ibid, at 53, 54–5.
119 Supra note 103, at 44
120 Ibid, at 43
121 Ibid, at 44–5. For merger of the Environment Co-ordination Board into the ACC machinery, see chapter 2 note 3. For discussion of the work of the committee which recommended this step, see 100–1 infra.
122 Supra note 113, at 45
123 Decision no. 7(II), *Report of the UNEP GC (2d session)*, supra note 3, at 86
124 U.N. Doc. UNEP/GC/30 (1975). See also *Report of the UNEP GC (3d session)*, supra note 3, at 22–5, 103–5. Cf Environment Programme (Levels One, and Three), U.N. Doc. UNEP/GC/90 (15 March 1977).
125 Establishment of a Marine Environment Protection Committee, IMCO Res. A.297 (VIII), in 13 *Int'l Legal Materials* 476, 477 (1974)
126 For discussion see Kiss, La protection de l'environnement et les organisations européennes, 19 *Annuaire français de droit international* 895 (1973).
127 Convention for the Prevention of Marine Pollution by Dumping from Ships and Aircraft, *done* 15 February 1972, in 11 *Int'l Legal Materials* 262, arts. 16–17 (1972); Convention for the Prevention of Marine Pollution from Land-Based Sources, *done* 4 June 1974, ibid, vol. 13, at 352, arts. 15–16 (1974). See Res.

248 Notes to chapter 4, pages 100–3

248 Notes to chapter 4, pages 100–3

no. 2, ibid, at 373, on the relationship between the two committees. See also chapter 3, 35–6, 37–8.

128 See Huntley, Man's Environment and the Atlantic Alliance 7 (NATO Info. Serv. 1971).

129 The reports of these groups' appraisals are respectively: Commission on International Development, *Partners in Development* (1969); *A Study of the Capacity of the United Nations Development System*, U.N. Doc. DP/5 (2 vols. 1969); and Conclusions and Recommendations of the Ad Hoc Committee on the Restructuring of the Economic and Social Sectors of the United Nations System, Annex to G.A. Res. 32/197 (9 January 1978), reprinted in 17 *Int'l Legal Materials* 235 (1978).

130 *Capacity Study*, supra note 129, foreword by Sir Robert Jackson, vol. 1, at iii

131 See G.A. Res. 32/197, supra note 29. For a consequence of this restructuring for UNEP and its Environment Co-ordination Board, see chapter 2, note 3, and 99 supra.

132 Res. 4(I), *Report*, supra note 1, at 67

133 G.A. Res. 2994, 27 U.N. GAOR Supp. 30, at 42, U.N. Doc. A/8730 (1972)

134 Decision No. 17(II), *Report of the UNEP GC (2d session)*, supra note 3, at 117; Decision No. 43(III), *Report of the UNEP GC (3d session)*, supra note 3, at 123

135 G.A. Res. 2997, supra note 113, at 44

136 Ibid

137 See 98 supra.

138 But the administrative costs of the Environment Fund, in addition to its operational program costs and program support, are borne by the Fund itself. These arrangements were a result of a compromise at Stockholm between those who wanted the Environment Programme to be fully supported by the regular UN budget and those who wanted it wholly financed by separate and voluntary financial arrangements.

139 See *Report*, supra note 1, at 110; see also ibid, at 63–4.

140 See Report on the Implementation of the Fund Programme in 1973, U.N. Doc. UNEP/GC/23, at 2 (1974); cf Review and Approval of the Fund Programme for 1974 and 1975, U.N. Doc. UNEP/GC/17/Rev.1, at 2–3 (1974).

141 See *Report of the UNEP GC (5th session)*, supra note 3, Decision 98(V) (B) at 146–7, and 81–9. See also Report on the Implementation of the Fund Programme in 1977, U.N. Doc. UNEP/GC.6/13 (1978) and its predecessor annual reports; Management of the Environment Fund, U.N. Doc. UNEP/GC.6/15 (1978).

142 OECD, Guiding Principles Concerning International Economic Aspects of Environmental Policies, in 11 *Int'l Legal Materials* 1172 (1972), reprinted in ibid,

vol. 14, at 236 (1975). See also OECD, Council Recommendation on Principles Concerning Transfrontier Pollution, ibid, vol. 14, at 242; OECD, *Problems in Transfrontier Pollution* (1974).

143 11 *Int'l Legal Materials*, at 1172–3
144 Ibid, at 1173
145 Ibid
146 OECD, Recommendation of the Council on the Implementation of the Polluter-Pays Principle, ibid, vol. 14, at 234 (1975). See also Note on the Implementation of the Polluter-Pays Principle, ibid, at 238.
147 European Communities, Council Recommendation on the Application of the Polluter-Pays Principle, ibid, vol. 14, at 138
148 See World Bank, *Environmental, Health, and Human Ecologic Considerations in Economic Development Projects* (1973). See generally International Institute for Environment and Development, Multilateral Aid and the Environment (1977); Schneider-Sawiris, Concept of Compensation in the Field of Trade and Environment (IUCN 1973).
149 See AID, Environmental Procedures, 22 CFR Part 216 (1978). For background considerations, see also AID, *Feasibility Studies, Economic and Technical Soundness Analyses, Capital Projects* (1964, addendum 1972); AID Manual Circulars nos. 1221.2: Consideration of Environmental Aspects of U.S.-Assisted Capital Projects (18 August 1970); and 1214.1: Procedure for Environmental Review of Capital Projects (20 September 1971). See generally IIED report, supra note 148.

CHAPTER FIVE

1 Beesley, The Canadian Approach – Environmental Law on the International Plane, in *Private Investors Abroad – Problems and Solutions in International Business in 1973*, at 239 (Proc. Southwestern Legal Fnda. 1973), reprinted in summary form in 11 *Can. Y.B. Int'l L.* 3 (1973)
2 Ambassador Beesley has several times presented this theory in the UN Sea-beds Committee and the present Law of the Sea Conference. The first extensive explanation of the concept seems to have come in a statement on 5 August 1971, and custodianship and delegation of powers were put forth in his remarks of 20 July 1972 and 9 March 1973. See also some references in various of his articles: supra note 1, at 256; The Arctic Pollution Prevention Act: Canada's Perspective, 1 *Syr. J. Int'l L. & Com.* 226, 233–4 (1973); The Law of the Sea Conference: Factors Behind Canada's Stance, *Int'l Perspectives* 28, 35 (July/August 1972).

3 Report of the Second Session of the Intergovernmental Working Group on Marine Pollution, U.N. Doc. A/Conf.48/IWGMP.II/5, at 12–13 (1971). For discussion, see chapter 3, 35 and 36.

4 Legault, Maritime Claims, in *Canadian Perspectives on International Law and Organization* 377, 392 (R. Macdonald, G. Morris, and D. Johnston eds 1974)

5 Ibid, at 393

6 See generally M. McDougal, H. Lasswell, and J. Miller, *The Interpretation of Agreements and World Public Order* (1967).

7 Legault, supra note 4, at 392–3

8 For further discussion see infra 'Funding,' 134–5.

9 *Report of the United Nations Conference on the Human Environment*, U.N. Doc. A/Conf.48/14, at 7 (1972) (hereinafter *Report*). Principle 21 is quoted in full at chapter 3, 21.

10 *Report* at 40. See also 117 infra.

11 G.A. Res. 31/72 Annex, 31 U.N. GAOR Supp. 39, at 36, U.N. Doc. A/31/39 (1976), reprinted in 16 *Int'l Legal Materials* 88 (1976); see chapter 3, 29.

12 *Done* 26 March 1975, [1975] 1 U.S.T. 540, T.I.A.S. no. 8056, art. 7

13 ICNT Art. 42(1) (b) and (2), in 8 *Third United Nations Conference on the Law of the Sea: Official Records* 11 (1977). During the negotiations on straits at UN-CLOS III, the United States, for example, proposed strict liability for accidents caused by failure to adhere to international pollution control standards and traffic separation schemes in straits used for international navigation (U.S.A.: Draft Articles on the Breadth of the Territorial Sea, Straits, and Fisheries, U.N. Doc. A/AC. 138/SC.II/L.4, reprinted in *Report of the Committee on the Peaceful Uses of the Sea-Bed and The Ocean Floor Beyond The Limits of National Jurisdiction*, 26 U.N. GAOR Supp. 21, at 241, U.N. Doc. A/8421 (1971)). For discussion, see Stevenson and Oxman, Preparations for the Law of the Sea Conference, 68 *Am. J. Int'l L.* 1, 10 (1974). Cf USSR: Draft Articles on Straits, U.N. Doc. A/AC.138/SC.II/L.7, in the next year's *Seabeds Report*, 27 U.N. GAOR Supp. 21, at 162, U.N. Doc. A/8721 (1972). There are very serious environmental drawbacks to such proposals – mainly that they do nothing about prevention and are addressed only to *ex post facto* monetary compensation – as straits states did not hesitate to point out. The current ICNT version, however, is not noticeably more helpful from either perspective.

14 See Coase, The Problem of Social Cost, 3 *J. L. & Econ.* 1 (1960); Calabresi, Transaction Costs, Resource Allocation and Liability Rules – A Comment, ibid, vol. 11, at 67 (1968).

15 *Report*, supra note 9, at 7, quoted in full at chapter 3, 48.

16 Supra note 10. See also Convention on the Prohibition of Military or Any Other Hostile Use of Environmental Modification Techniques, supra note 11, discussed

at chapter 3, 29; US and Soviet Offer a Pact on Weather, *N.Y. Times*, 22 August 1975, at 3, col. 1.

17 Supra note 9

18 G.A. Res. 3129, 28 U.N. GAOR Supp. 30, at 48, U.N. Doc. A/9030 (1973); Principles Concerning Transfrontier Pollution, OECD Doc. C(74)224 (21 November 1974), reprinted in 14 *Int'l Legal Materials* 242 (1975); ICNT arts. 199, 207, in 8 *Third UNCLOS: Official Records*, supra note 13, at 35, 36. At the seventh session of UNCLOS III, states also agreed on an additional notification provision to be added to ICNT art. 212, ibid, at 37. See Results of Negotiations on Part XII, UNCLOS Doc. MP/24, at 1 (15 May 1978).

19 See generally the discussion of 'draft Principle 20' and subsequent events at chapter 3, 51.

20 See G.A. Res. 3129, supra note 18; G.A. Res. 3281, 29 U.N. GAOR Supp. 31, at 50, U.N. Doc. A/9631 (1974). For some early accomplishments in fulfilment of this mandate, see Report of the Group of Experts on Liability for Pollution and Other Environmental Damage and Compensation for Such Damage, U.N. Doc. UNEP/WG.8/3 (6 April 1977); Report of the Working Group of Experts on Environmental Law on Its Second Session, U.N. Doc. UNEP/WG.14/4 (12 April 1978); Report of the Intergovernmental Working Group of Experts on Natural Resources Shared by Two or More States on the Work of Its Fifth Session, U.N. Doc. UNEP/IG.12/2 (8 February 1978).

21 *Done* 4 June 1974, in 13 *Int'l Legal Materials* 352 (1974), discussed at chapter 3, 37–8. See also UNEP Report on Draft Protocol for the Protection of the Mediterranean Sea against Pollution from Land-Based Sources, 16 *Int'l Legal Materials* 958 (1977).

22 *Done* 29 April 1958, [1966] 1 U.S.T. 138, T.I.A.S. no. 5969, 559 U.N.T.S. 285; see chapter 3, 55–6.

23 See chapter 3, 21 and 211 n19.

24 ICNT art. 209, in 8 *Third UNCLOS: Official Records*, supra note 13, at 36, it should be noted, does commit coastal states to 'establish national laws and regulations to prevent, reduce and control pollution of the marine environment arising from or in connexion with sea-bed activities subject to their jurisdiction'; it also allows them to 'take other measures as may be necessary to prevent, reduce and control such pollution.' As to flag states, in Annex II para. 15, ibid, at 54, there is the rather curious savings clause that 'the application by a State Party of environmental regulations to sea-bed miners it sponsors or to ships flying its flag, more stringent than those imposed by the Authority ... shall not be deemed inconsistent with ... the present Convention'; while sought by environmentalists and therefore in some sense a political victory for them, this last provision could be somewhat misleading, as nowhere else in the ICNT is

the authority of a flag state to impose whatever additional standards it may wish (employment, tax, construction and design, and so forth) even brought into question in equivalent manner. For discussion, see Schneider, Something Old, Something New: Some Thoughts on Grotius and the Marine Environment, *Va. J. Int'l L.* 147, 163 (1977). On one type of sea-bed-source pollution, see Dubais, Compensation for Oil Pollution Damage Resulting from Exploration and Exploitation of Hydrocarbons in the Seabed, 6 *J. Maritime L. & Com.* 549 (1974).

25 This effect was, in fact, long ago suggested by David Hume: 'There is no quality in human nature which causes more fatal errors in our conduct, than that which leads us to prefer whatever is present to the distant and remote, and makes us desire objects more according to their situation than their intrinsic value. Two neighbours may agree to drain a meadow, which they possess in common: because it is easy for them to know each other's mind; and each must perceive, that the immediate consequence of his failing his part, is the abandoning of the whole project. But it is very difficult, and indeed impossible, that a thousand persons should agree in any such action; it being difficult for them to concert so complicated a design, and still more difficult for them to execute it; while each seeks a pretext to free himself of the trouble and expense, and would lay the whole burden on others' (2 D. Hume, *A Treatise of Human Nature* 239 (Everyman ed 1952)).

26 See, eg, Environmental Law Institute, *Federal Environmental Law* 492–545 (E. Dolgin and T. Guilbert eds 1974); O. Gray, *Cases and Materials on Environmental Law* 1001–78 (1970); E. Hanks, A. Tarlock, and J. Hanks, *Cases and Materials on Environmental Law and Policy* 457–706 (1974); F. Grad, *Environmental Law: Sources and Problems*, chapters 8–9 (1971). See generally D. Hagman, *Public Planning and Control of Urban and Land Development* (1973). But see B. Ackerman and S. Ackerman, J. Sawyer, and D. Henderson, *The Uncertain Search for Environmental Quality* (1974).

27 *Report of the World Population Conference*, U.N. Doc. E/Conf.60/19, at 17 (1974)

28 Ibid, at 7, 25

29 Trade Act of 1974, 19 U.S.C. § 2101, 2432 (1976), reprinted in 14 *Int'l Legal Materials* 181, § 402 (1975). An exchange of letters concerning emigration from the USSR between the United States and that country appears in 14 *Int'l Legal Materials* at 248.

30 See 122 supra; see also chapter 3, 73.

31 See *Report of the World Food Conference*, U.N. Doc. E/5587 (1974).

32 For discussion, see, eg, Choucri and North, Dynamics of International Conflict: Some Policy Implications of Population, Resources and Technology, 24 *World Pol.* 80 (Supp. 1972).

33 On this subject see generally U.N. Fund for Population Activities, Law and World Population, U.N. Doc. E/Conf.60/BP/6 (1974).
34 See ibid, at 23–5.
35 See McDougal, Lasswell, and Chen, Human Rights for Women and World Public Order: The Outlawing of Sex-Based Discrimination, 69 *Am. J. Int'l L.* 497 (1975).
36 For a short discussion of this topic, see, eg, Montgomery, The Case for Compulsory Regulation of Reproduction, in *The World Population Crisis: Policy Implications and the Role of Law* 67 (Proc. Am. Soc. Int'l L. regional meeting and John Bassett Moore Soc. Int'l L. symposium 1971).
37 See chapter 4, 76–8.
38 See, eg, B. Russett and H. Alker, K. Deutsch, and H. Lasswell (1st ed) and C. Taylor and M. Hudson (2d ed), *World Handbook of Political and Social Indicators* (1964 and 1972 respectively).
39 D. Meadows, D. Meadows, J. Randers, and W. Behrens, *The Limits to Growth* (1972); compare M. Mesarovic and E. Pestel, *Mankind at the Turning Point* (1974); see generally The No-Growth Society, 102 *Daedalus* (Fall 1972).
40 The basic terms of the consent regime for marine scientific research (msr) in the zone are set forth in ICNT art. 247, in 8 *Third UNCLOS: Official Records*, supra note 13, at 42. On msr generally, see arts. 239–66, ibid, at 41–4.
41 ICNT arts. 205–6, ibid, at 36. See also art. 207 on assessment of potential effects of activities on the marine environment, ibid.
42 ICNT art. 65, ibid, at 15. See also art. 120 on conservation and management of marine mammals on the high seas, ibid, at 21.
    For the original moratorium at the UNCHE, see *Report*, supra note 9, at 23. The moratorium was subsequently rejected at the next session of the International Whaling Commission one month later, and it has since only been accepted and lived up to in limited form. See also *Report of the Governing Council of the United Nations Environment Programme on the Work of Its Third Session*, 30 U.N. GAOR Supp. 25, at 112, Decision 33(III), U.N. Doc. A/10025 (1975).
43 This was one of the important conclusions of the August 1973 Colloquium of the Academy of International Law of the Hague on the Protection of the Environment and International Law.
44 *Report of the Governing Council of the United Nations Environment Programme on the Work of Its Second Session*, 29 U.N. GAOR Supp. 25, at 107, Decision 11(II), U.N. Doc. A/9625 (1974). See also Revolving Fund (Information), U.N. Doc. UNEP/GC/47 (1975).
45 See, eg, Rules of Procedure of the Third U.N. Conference on the Law of the Sea, U.N. Doc. A/Conf.62/30/Rev.1 (1974), which provide for voting only upon prolonged inability to achieve consensus.

46 See particularly *Report of the ILC*, 11 U.N. GAOR Supp. 9, U.N. Doc. A/3159 (1956), reprinted in 51 *Am. J. Int'l L.* 154 (1957).

47 *Reports of the UNEP GC (2d & 3d sessions)*, supra notes 44 and 42, at 101 and 114, Decisions 8(II) and 35(III) respectively. For some early accomplishments pursuant to these mandates, see note 20 supra.

48 The International Convention Relating to Intervention on the High Seas in Cases of Oil Pollution Casualties and the International Convention on Civil Liability for Oil Pollution Damage, *done* 29 November 1969, in 9 *Int'l Legal Materials* 25 and 45 (1970) came into force on 6 May and 19 June 1975 respectively; a Protocol to the former, *done* 2 November 1973, appears in ibid, vol. 13, at 605 (1974). The Convention *done* 12 May 1954, [1961] 3 U.S.T. 2989, T.I.A.S. no. 4900, 327 U.N.T.S. 3, as amended 21 October 1969, appears in 9 *Int'l Legal Materials* 1 (1970); the 1969 amendments entered into force on 20 January 1978, T.I.A.S. no. 8505.

49 The 1971 amendments to the 1954 convention, supra note 48, appear in 11 *Int'l Legal Materials* 267 (1972). The 1971 Fund Convention appears in ibid, at 284, and the 1973 Pollution from Ships Convention in ibid, vol. 12 at 1319 (1973). See generally Status of International Conventions relating to Marine Pollution of which IMCO Is Depository or Is Responsible for Secretariat Duties, IMCO Doc. MEPC IX/2 (17 March 1978).

50 Supra note 48, arts. 16 and 3 respectively. 'Tacit amending' provisions have also, for example, been written into the 1974 International Convention on the Safety of Life at Sea ('SOLAS' Convention), *done* 1 November 1974, art. 8, in 14 *Int'l Legal Materials* 959 (1975).

51 As has frequently been pointed out by the jurists, the question of the *locus standi* of states in this regard remains in doubt. Compare *South West Africa* cases, Preliminary Objections (*Ethiopia* v *South Africa*; *Liberia* v *South Africa*), [1962] I.C.J. 319, with *South West Africa* cases, Second Phase, [1966] I.C.J. 6, between which a change in the composition of the Court resulted in the reversal of majority and minority positions; see also *Northern Cameroons* case, Preliminary Objections (*Cameroon* v *United Kingdom*), [1963] I.C.J. 15. For discussion, see I. Brownlie, *Principles of Public International Law* 453–9 (2d ed 1973).

52 [1974] I.C.J. 253 and 457. For discussion, see chapter 3, 41–2, chapter 4, 91–2.

53 See Gardner, The Role of the UN in Environmental Problems, 26 *Int'l Org.* 237, 254 (1972).

54 Convention on the Protection of the Environment, *done* 19 February 1974, in 13 *Int'l Legal Materials* 591 (1974). The parties are Denmark, Finland, Norway, and Sweden. See chapter 3, 53, chapter 4, 91.

55 See chapter 4, 85–6, 94; see also chapter 6, 145–6.

56 See 130 and note 52 supra.

57 Former I.C.J. Judge Philip Jessup has presented these and other suggestions in Do New Problems Need New Courts? 65 *Proc. Am. Soc. Int'l L.* 261 (1971).

58 See Principles 8–14 and Recommendations 102–9, *Report* supra note 9, at 5 and 54–8. See also *Development and Environment* 6 and passim (the 'Founex Report' 1971).

59 For discussion see chapter 3, 69; see also chapter 4, 97–8.

60 S. 2801 and H.R. 13904, 92d Cong., 2d Sess. (1972); reintroduced as S. 1134 and H.R. 9, 93d Cong., 1st Sess. (1973); most recently S. 2053, 95th Cong., 1st Sess. (1977). The 'Metcalf Bill' is discussed as part of the broad question of unilateral state action in chapter 4, 86–7.

61 On this subject see Strong, One Year After Stockholm: An Ecological Approach to Management, 51 *Foreign Aff.* 691 (1973); see generally *Managing the Planet* (P. Albertson and M. Barnett eds 1972); *Public Policy Toward Environment 1973: A Review and Appraisal* (Annals N.Y. Acad. Sci. vol. 216, 1973).

62 This is in accord with Decisions 17(II) and 43(III) of the UNEP Governing Council on the question of convening a second UN Conference on the Human Environment (*Reports of the UNEP GC (2d & 3d sessions)*, supra notes 44 and 42, at 117 and 123). See also Question of Convening a Second United Nations Conference on the Human Environment, U.N. Doc. UNEP/GC/43 (1975).

63 *Report*, supra note 9, at 110

64 See, eg, *Explorations in the Theory of Anarchy* (G. Tullock ed 1972).

65 R. Falk, *This Endangered Planet* (1971) to some extent represents this perspective, and it is central to several of the 'world federalist' arguments; see chapter 1 n4. W. Ophuls, Prologue to a Political Theory of the Steady State (unpub. dissertation in Sterling lib. at Yale 1973) has done a whole thesis essentially from this point of view.

66 See, eg, Canada, Fiji, Ghana, Guyana, Iceland, India, Iran, New Zealand, Philippines, and Spain, Draft Articles on a Zonal Approach to the Preservation of the Marine Environment, U.N. Doc. A/Conf.62/C.3/L.6 (1974), reprinted in 3 *Third United Nations Conference on the Law of the Sea: Official Records* 249 (1975).

67 For discussion of the incident and search, see Cosmos 954: An Ugly Death, *Time* 28 (6 February 1978). On overall costs, see, eg, Wilford, Canadians End Search for Debris of Soviet Satellite, *N.Y. Times*, 2 April 1978, at 10, col. 1.

CHAPTER SIX

1 Treaty of Guadelupe-Hidalgo – International Law, 21 *Op. Att'y Gen.* 274, 281, and 283 (1893–7). For some discussion of this matter see Utton, International

Environmental Law and Consultation Mechanisms, 12 *Colum. J. Transnat'l L.* 56, 57–9 (1973).

2 I. Brownlie, *Principles of Public International Law* 418–19 (2d ed 1973)

3 Convention on the International Responsibility of States for Injuries to Aliens 45 (Draft no. 12, with Explanatory Notes, L. Sohn and R. Baxter reporters 1961). See generally C. Amerasinghe, *State Responsibility for Injuries to Aliens* (1967).

4 On this subject of who is responsible for what, see Fatouros, Developing Legal Standards of Liability for Transnational Environmental Injury: Bases of Liability and Standing to Complain, in *International Responsibility for Environmental Injury* (R. Stein ed forthcoming). And there are also questions of agency and joint tortfeasors. See Brownlie, supra note 2, at 441–4.

5 See Brownlie, ibid.

6 *Report of the International Law Commission on the Work of Its Twenty-Ninth Session*, U.N. Doc. A/32/283, art. 19(3)(d) (1977). The ILC has been giving attention at this stage of its deliberations to the responsibility of states for international wrongful acts, leaving aside the problem of responsibility for risks. After it completes work on these draft articles, the Commission has been given a mandate by the General Assembly to turn to the topic of international liability for injurious consequences arising out of acts not prohibited by international law (G.A. Res. 3315, 29 U.N. GAOR Supp. 31, at 144, U.N. Doc. A/9631 (1975)).

7 *Report of the United Nations Conference on the Human Environment*, U.N. Doc. A/Conf.48/14, at 7 (1972) (hereinafter *Report*), quoted previously at chapter 3, 21 and 48

8 R.S.C. 1970 (1st Supp.) c. 2 (1970); for discussion see chapter 3, 27, chapter 4, 81, 85–6.

9 The basic political ecological problem is as follows. On the one hand, claims of sovereignty on a sector theory have been raised as to certain parts of the Arctic by the Soviet Union. But on the other, the United States, among others, has strongly and consistently opposed the advancement of any such claims or the drawing of baselines by Canada in its neighbouring northern areas. As a result, any determination in an international forum of rules and regulations to apply to the Arctic, which would in effect refer to the Arctic north of Canada overwhelmingly, is not politically viable from the Canadian point of view. See generally D. Pharand, *The Law of the Sea in the Arctic* (1973).

10 Supra note 8, preamble

11 Ports and Waterways Safety Act of 1972, 33 U.S.C.A. § 1221 *et seq.* (Supp. V 1975); for discussion, see chapter 4, 86, 94.

12 See chapter 4, 93–4 *et seq.*

13 33 U.S.C. § 1227

14 Reference is made to the Environmental Law Information System and the Environmental Law Centre of the International Union for the Conservation of Nature and Natural Resources (IUCN) at chapter 4, 78 and chapter 5, 126.
15 See *Black's Law Dictionary* 1507 (rev 4th ed 1968).
16 On state responsibility and the role of declaratory judgments see the report by F.V. García-Amador, Special Rapporteur, [1961] 2 *Y.B. Int'l L. Comm'n* 1, 14–16, U.N. Doc. A/CN.4/134 (1961). He explains that sometimes declaratory judgments 'constitute a simple means of giving satisfaction for "moral and political" injury caused to a State, or, in other words, a method of "making reparation" for an act contrary to international law by formally declaring it to be unlawful and thus sanctioning or censuring the conduct imputable to the defendant state'; at other times such a judgment 'constitutes a type of "juridical reparation" for the unlawfulness of an act or omission capable of occasioning actual and effective injury and therefore constitutes a form of reparation *sui generis*' (ibid, at 15–16).
17 The International Court of Justice stated that 'to ensure respect for international law ... the Court must declare that the action of the British Navy constituted a violation of Albanian sovereignty' and that '[t]his declaration is in accordance with the request made by Albania through her Counsel [for "the declaration of the Court from a legal point of view"] and is in itself appropriate satisfaction' ([1949] I.C.J. 4, 35). See chapter 3, 49.
18 Handl, Territorial Sovereignty and the Problem of Transnational Pollution, 69 *Am. J. Int'l L.* 50, 72–5 (1975). But see the second use of declaratory judgments described in García-Amador's report, supra note 16.
19 (*United States* v *Canada*), 3 U.N.R.I.A.A. 1938, 1963 (1941). For discussion see chapter 3, 48–9; see also García-Amador, supra note 16, at 15.
20 *Nuclear Tests (Australia* v *France*) and *Nuclear Tests (New Zealand* v *France*), [1974] I.C.J. 253 and 457. For discussion see chapter 3, 41–2, chapter 4, 91–2. See also Goldie, The Nuclear Tests Cases: Restraints on Environmental Harm, 5 *J. Maritime L. & Com.* 491 (1974); Frank, Word Made Law: The Decision of the International Court of Justice in the Nuclear Test Cases, 69 *Am.J. Int'l L.* 612 (1975).
21 *Nuclear Tests* cases, supra note 20, at 263 and 457
22 Ibid, at 272. It consequently also found that the application of Fiji for permission to intervene lapsed and that therefore 'no further action thereon is called for on the part of the Court' (order of 20 December 1974, [1974] I.C.J. 530, 531).
23 See Joint Dissenting Opinion of Judges Onyeama, Dillard, Jiménez de Aréchaga, and Sir Humphrey Waldock, [1974] I.C.J. 494.
24 General Principles for the Assessment and Control of Marine Pollution, *Report*, supra note 7, Annex III, at 1. For the genesis of the General Principles and their accompanying Statement of Objectives, see chapter 3, 35.

25 Convention for the Prevention of Marine Pollution from Land-Based Sources, *done* 4 June 1974, in 13 *Int'l Legal Materials* 352, art. 12 (1974); see chapter 3, 37–8.

26 Convention for the Prevention of Marine Pollution by Dumping from Ships and Aircraft, *done* 15 February 1972, in 11 *Int'l Legal Materials* at 262 (1972) and Convention on Prevention of Marine Pollution by Dumping of Wastes and Other Matter, *done* 29 December 1972, [1975] 2 U.S.T. 2403, T.I.A.S. no. 8165; see chapter 3, 35–6.

27 For discussion and citations see chapter 3, 32–3, 35–6, 36, 36–7, 37–8, 39, 40, 40–1, 42–3, and 29 respectively. See also Legault, The Freedom of the Seas: A Licence to Pollute? 21 *U. Toronto L.J.* 211, 216 (1971).

28 See chapter 3, 36, chapter 4, 94.

29 For further specification, see the enforcement provisions of the Informal Composite Negotiating Text (ICNT), arts. 214–23, in 8 *Third United Nations Conference on the Law of the Sea: Official Records* 37–9 (1977). See also the safeguards in ICNT arts. 224–34, ibid, at 39–40.

30 See *Report*, supra note 7, at 59.

31 For description see chapter 4, 76–7. See generally Global Environmental Monitoring System, U.N. Doc. UNEP/GC/31/Add.2 (1975); Development and Implementation of the Global Environmental Monitoring System, U.N. Doc. UNEP/GC/Inf.2 (1977); Progress Report on International Referral System Development, U.N. Doc. UNEP/GC/Inf.7 (1977).

32 *Report of the Governing Council of the United Nations Environment Programme on the Work of Its Fifth Session*, 32 U.N. GAOR Supp. 25, at 9, U.N. Doc. A/32/25 (1977). See also Decision 84(V)(B), ibid, at 122.

33 Results of Consideration of Proposals and Amendments Relating to the Preservation of the Marine Environment, U.N. Doc. A/Conf.62/C.3/L.15/Add.1, at 1, art. 8 (1975). As originally agreed to, the draft article reads as quoted in the text. In accordance with a subsequent trend to refer only to 'competent international organizations' at the present stage of the negotiations, the article now commits states to transmit their reports 'to the competent international or regional organizations, which shall make them available to States' (ICNT art. 206, in 8 *Third UNCLOS: Official Records*, supra note 13, at 36, quoted 162 infra). When the question of institutional roles is finally dealt with, nevertheless, it is fully expected that UNEP will be recognized as a competent and appropriate organization in this area.

34 For presentation of the facts and analysis of the legal issues raised by this disaster, see Brown, The Lessons of the *Torrey Canyon*, 21 *Current Legal Problems* 113 (1968); McGurren, The Externalities of a *Torrey Canyon* Situation: An Impetus for Change in Legislation, 11 *Natural Resources J.* 349 (1971);

Utton, Protective Measures and the *Torrey Canyon*, 9 *B.C. Ind. & Com. L. Rev.* 613 (1968); Comment, Post *Torrey Canyon*: Toward a New Solution to the Problem of Traumatic Oil Spill, 2 *Conn. L. Rev.* 632 (1970).

35 For details see Canada Asks US Payment for Oil Spill on West Coast, *N.Y. Times*, 10 June 1972, at 36, col. 5; Oil in Canadian Waters (edit.), ibid, 27 June 1972, at 40, col. 1.

36 Statement on the Cherry Point Oil Spill by Hon. Mitchell Sharp, then Secretary of State for External Affairs, House of Commons Debates, quoted in 11 *Can. Y.B. Int'l L.* 314, 333–4 (1973). On the Canadian reaction in general and subsequent handling of clean-up costs, see: US Oil Spill Leaving Film on BC Beach, *Gazette* (Montreal), 6 June 1972; Municipal Workers, Volunteers Cleaning Up BC Spill, *Globe & Mail* (Toronto), 7 June 1972; BC Oil Spill Tars Pipeline's Future, *Journal* (Ottawa), 7 June 1972; MPs Back Demand for Oil Spill Payment, *Montreal Star*, 9 June 1972; Compensation Demanded for Oil Spill in BC, *Evening Telegram* (St John's), 9 June 1972; Sellar, Oil Spill Damages Sought, *Calgary Herald*, 9 June 1972; US Oil Company Billed for Spill, *Montreal Star*, 29 June 1972; ARCO Sets Up Offices for Oil Damage Bills, *Columbian* (New Westminster), 24 June 1972; ARCO Pays $19,000 Cleanup Bill, *Sun* (Vancouver), 10 August 1972; Oil Spill Billing Will Hit $26,000, *Columbian* (New Westminster), 8 August 1972.

37 For facts and discussion of the *Metula*, see Third Committee, Summary Record of the Fifteenth Meeting, U.N. Doc. A/Conf.62/C.3/SR.15 (1974), reprinted in 2 *Third UNCLOS: Official Records* 372, 373–5 (1975). On subsequent tanker accidents, see McManus and Schneider, Shipwrecks, Pollution & the Law of the Sea, *Nat'l Parks & Conservation Magazine* 10 (June 1977). For discussion of the *Amoco Cadiz*, see Third Committee, Provisional Summary Record of the 35th Meeting, U.N. Doc. A/Conf.62/C.3/SR.35 (Prov. 20 April 1978).

38 Mostert, Profiles: Supertankers, *New Yorker* 45 and 46 (2 parts, 13 and 20 May 1974), the quotation being from part 2, at 75. See also N. Mostert, *Supership* (1974), which expands upon these articles. On cleaning up these spills, see generally Environmental Emergency Branch, Environmental Protection Service, Canada, *Spill Technology Newsletter* (bimonthly).

39 See generally F. Lewis, *One of Our H-Bombs is Missing* (1967); T. Szulc, *The Bombs of Palomares* (1967).

40 Radioactive Spanish Earth Is Buried 10 Feet Deep in South Carolina, *N.Y. Times*, 12 April 1966, at 28, col. 3

41 On these final details, see Szulc, H-Bomb is Recovered After 80 Days, ibid, 8 April 1966, at 1, col. 4; Szulc, Dented H-Bomb Is Displayed on Recovery Ship, ibid, 9 April 1966, at 1, col. 4.

42 See the discussion of self-defence, self-preservation, and security in M. McDougal and F. Feliciano, *Law and Minimum World Public Order* 213–16 et seq. (1961); see also chapter 3, 46–7.
43 *Done* 29 November 1969, [1975] 1 U.S.T. 765, T.I.A.S. no. 8068, discussed at chapter 3, 33–4.
44 The plan itself was adopted on 10 June 1971 and is reprinted in 6 *New Directions in the Law of the Sea* 464 (S. Lay, R. Churchill, and M. Nordquist, eds 1975). It was described in and its purpose is here cited from the Agreement on Great Lakes Water Quality, 15 April 1972, [1972] 1 U.S.T. 301, T.I.A.S. no. 7312, Annex 8.
45 *Entered into force* 9 August 1969, in 9 *Int'l Legal Materials* 359, art. 6 (1970). For discussion of other aspects of the arrangement see chapter 3, 34–5.
46 9 *Int'l Legal Materials* 359, art. 1
47 *Done* 8 February 1949, [1950] 1 U.S.T. 477, T.I.A.S. no. 2089, 157 U.N.T.S. 157. The workings of ICNAF were described in chapter 3, 57.
48 On the general subject of the international legal implications, see Stein, Cannikin, in *International Responsibility for Environmental Injury*, supra note 4. On the protests see, eg, Kentworthy, Nixon May Cancel Aleutians A-Test, *N.Y. Times* 9 September 1971, at 1, col. 4; Szulc, US, Britain Wary on Security Talks, ibid, 2 October 1971, at 1, col. 7; Walz, 12 Sail for Amchitka to Fight Atom Test, ibid, 3 October 1971, at 14, col. 4; Gamble in the Aleutians (edit.), ibid, 4 October 1971, at 38, col. 2; Kentworthy, Nixon Authorizes Atomic Explosion in the Aleutians, ibid, 28 October 1971, at 1, col. 4; Canada Voices Disquiet, ibid, 28 October 1971, at 26, col. 8; Turner, H-Bombs Tested in the Aleutians Despite Protest, ibid, 7 November 1971, at 1, col. 8; Shock Waves Felt in Japan, ibid, 7 November 1971, at 64, col. 6.
49 On the legal issues involved see Weiss, Project West Ford: Needles in Space, in *International Responsibility for Environmental Injury*, supra note 4. On the protests see, eg, Finney, Needle Antennas Stir Space Furor, *N.Y. Times*, 30 July 1961, at 48, col. 1; Sullivan, Needles Orbiting to Be Tried Again, ibid, 3 February 1962, at 5, col. 1; Finney, New Panel to Screen Space Experiments, ibid, 10 May 1962, at 16, col. 4; Needles Orbited for Radio Relay, ibid, 13 May 1963, at 1, col. 5; US Assures World Scientists Needles Are Harmless, ibid, 18 May 1963, at 9, col. 7; Soviet Again Hits US Needles Test, ibid, 21 May 1963, at 3, col. 1; Brewer, Soviet Is Accused of Space Secrecy, ibid, 7 June 1963, at 10, col. 3; More Needles in Space? (edit.), ibid, 23 September 1963, at 28, col. 2.
50 Remarks of Mr Kearney, [1973] 1 *Y.B. Int'l L. Comm'n* 7
51 See *Committee for Nuclear Responsibility* v *Seaborg*, 463 F.2d 783 (D.C. Cir. 1971).
52 *Committee for Nuclear Responsibility* v *Seaborg*, 463 F.2d 788 (D.C. Cir. 1971)
53 *Committee for Nuclear Responsibility* v *Seaborg*, 463 F.2d 796 (D.C. Cir. 1971)

54 *Committee for Nuclear Responsibility* v *Schlesinger*, 404 U.S. 917, 92 S. Ct. 242, 30 L. Ed. 2d 191 (1971)
55 See Turner, H-Bombs Tested in the Aleutians Despite Protest, supra note 48.
56 See Mr Justice Douglas' dissent, 404 U.S. at 917; Justices Brennan and Marshall would also have granted a temporary restraining order pending plaintiff's filing of a petition for certiorari and action by the Court on the petition (404 U.S. at 930).
57 *Wilderness Soc'y* v *Hickel*, 325 F. Supp. 422, 424 (D.D.C. 1970). See also *Natural Resources Defense Council* v *Morton*, 458 F.2d 827 (D.C. Cir. 1972).
58 See *Wilderness Soc'y* v *Morton*, 479 F.2d 842 (D.C. Cir. 1973).
59 *Fisheries Jurisdiction (United Kingdom* v *Iceland)* and *Fisheries Jurisdiction (Federal Republic of Germany* v *Iceland)*, Interim Protection, Orders of 17 August 1972, [1972] I.C.J. 12, 17 and 30, 35. For discussion of these cases, see chapter 3, 58–9.
60 Interim Protection Orders, [1972] I.C.J. at 17 and 35
61 Jurisdiction of Court, Judgments, [1973] I.C.J. 3 and 49
62 Interim Measures, Orders of 12 July 1973, [1973] I.C.J. 302 and 313
63 Dissenting Opinion of Judge Petrén, [1973] I.C.J. 310. The other is the Dissenting Opinion of Judge Gros, [1973] I.C.J. 306.
64 Declaration of Judge Ignacio-Pinto, [1973] I.C.J. 304, 305
65 *Done* 13 November 1973, in 12 *Int'l Legal Materials* 1315 (1973). See also Belgium-Iceland: Agreement on Fishing Within Fifty Mile Limit Off Iceland, *done* 7 September 1972, ibid, vol. 11, at 941 (1972); Iceland-Norway: Agreement Concerning Fishing Rights, *done* 10 July 1973, ibid, vol. 12, at 1313 (1973). Compare Belgium-Iceland: Fisheries Agreement on the Extension of the Icelandic Fishery Limits to 200 Miles, 28 November 1975, ibid, vol. 15, at 1 (1976); Federal Republic of Germany-Iceland: Fisheries Agreement on the Extension of the Icelandic Fishery Limits to 200 Miles, 28 November 1975, ibid, at 43; Iceland-Norway: Agreement concerning Norwegian Fishing in Icelandic Waters, 10 March 1975, ibid, at 875; Iceland-United Kingdom: Agreement concerning British Fishing in Icelandic Waters, 1 June 1976, ibid, at 878.
66 Merits, Judgments of 25 July 1974, [1974] I.C.J. 3 and 175
67 *Nuclear Tests (Australia* v *France)* and *Nuclear Tests (New Zealand* v *France)*, Interim Protection, Orders of 22 June 1973, [1973] I.C.J. 99, 106 and 135, 142. See 147 supra; see also chapter 3, 41–2, chapter 4, 91–2.
68 Draft Declaration on the Human Environment, U.N. Doc. A/Conf.48/4, Annex, para. 20, at 4 (1972), previously quoted at chapter 3, 51. For discussion see Sohn, The Stockholm Declaration on the Human Environment, 14 *Harv. Int'l L.J.* 423, 496–502 (1973).
69 Sohn, supra note 68, at 497, quoting U.N. Doc. A/Conf.48/PC.12, at 8 (1971)

70 Ibid, quoting U.N. Doc. A/Conf.48/CRP.5 (1972)

71 Ibid, quoting U.N. Doc. A/Conf.48/14, at 119 (1972)

72 G.A. Res. 2995, 27 U.N. GAOR Supp. 30, at 42, U.N. Doc. A/8730 (1972)

73 Ibid

74 G.A. Res. 2996, 27 U.N. GAOR Supp. 30, at 42, 43, U.N. Doc. A/8730 (1972)

75 G.A. Res. 3129, 28 U.N. GAOR Supp. 30, at 48, 49, U.N. Doc. A/9030 (1973)

76 *Report of the Governing Council of the United Nations Environment Programme on the Work of Its Second Session,* 29 U.N. GAOR Supp. 25, at 119, Decision 18(II), U.N. Doc. A/9625 (1974). See also ibid, at 54–64. This decision was taken by a vote of twenty-nine in favour to one against and sixteen abstentions, and Brazil cast the only negative vote (ibid, at 61).

77 Co-operation in the Field of the Environment Concerning Natural Resources Shared by Two or More States: Report of the Executive Director, U.N. Doc. UNEP/GC/44, at 41 (1975)

78 *Report of the Governing Council of the United Nations Environment Programme on the Work of Its Third Session,* 30 U.N. GAOR Supp. 25, at 124, Decision 44 (III), U.N. Doc. A/10025 (1975). See also ibid, at 82–6.

79 The Intergovernmental Working Group of Experts on Natural Resources Shared by Two or More States held its first session in Nairobi in January 1976. It met four more times within the next two years. The reports of these five sessions are respectively U.N. Docs. UNEP/GC/74 (1976), UNEP/IG.3/3 (1976), UNEP/IG.7/3 (1977), UNEP/IG10/2 (1977), and UNEP/IG.12/2 (1978).

80 OECD Doc. C(74)224 (21 November 1974), reprinted in 14 *Int'l Legal Materials* 242, 246, Titles G and F respectively (1975). Also relevant is Title E on the principle of information exchange and consultation. See also Council Recommendation on Implementing a Regime of Equal Right of Access and Non-Discrimination in relation to Transfrontier Pollution, 17 May 1977, ibid, vol. 16, at 977 (1977). See chapter 3, 51–2.

81 *Done* 26 March 1975, [1975] 1 U.S.T. 540, T.I.A.S. no. 8056, art. 3

82 Ibid, the quotation being from art. 5 and the emergency exemption found in art. 6

83 ICNT arts. 207, 206, in 8 *Third UNCLOS: Official Records,* supra note 29, at 36. Moreover, this says nothing about a duty to consult. It also appears that, unfortunately, a draft article which used to provide that coastal states could permit dumping in their zones only 'after due consultation' with other States that might be affected will now apparently allow them to do so simply 'after due consideration' of the rights of others (ibid, art. 211(5), as emerging from the Seventh Session of UNCLOS III).

84 ICNT art. 199, in ibid, at 35. It is worth noting in passing that warning and consultation provisions have already been written into the London Ocean Dump-

ing Convention, supra note 26, art. 5, and into the regional Agreement Concerning Pollution of the North Sea by Oil, supra note 45, art. 6.

85 Supra 143–4
86 See M. McDougal and F. Feliciano, supra note 42, at 287–96; M. McDougal, H. Lasswell and I. Vlasic, *Law and Public Order in Space* 404–6 *et seq.* (1963).
87 For a good summary treatment of principles and problems of international responsibility of states and international claims, see W. Bishop, *International Law* 742–899 (3d ed 1962).
88 See generally Goldie, Liability for Damage and the Progressive Development of International Law, 14 *Int'l & Comp. L.Q.* 1189 (1965); W. Jenks, *The Prospects of International Adjudication* 514–46 (1964).
89 Goldie, International Principles of Responsibility for Pollution, 9 *Colum. J. Transnat'l L.* 283, 306 (1970)
90 L.R. 1 H.L. 330 (1868), extracted in H. Shulman and F. James, *Cases and Materials on the Law of Torts* 61 (2d ed 1952)
91 H. Shulman and F. James, supra note 90, at 70. And as Lord Cranworth added: 'If a person brings, or accumulates, on his land anything which, if it should escape, may cause damage to his neighbour, he does so at his peril. If it does escape, and cause damage, he is responsible, however careful he may have been, and whatever precautions he may have taken to prevent the damage' (ibid, at 71). See also Bohlen, The Rule in *Rylands* v *Fletcher*, 59 *U. Pa. L. Rev.* 298 (1911).
92 (*United States* v *Canada*), 3 U.N.R.I.A.A. 1911 and 1938 (1938 and 1941), reprinted in 33 *Am. J. Int'l L.* 182 (1939) and ibid, vol. 35, at 684 (1941). See chapter 3, 48–9.
93 Ibid, at 1965; see chapter 3, 48–9.
94 See Canada-United States Settlement of Gut Dam Claims, 22 September 1968, Report of the Agent of the United States before the Lake Ontario Claims Tribunal, in 8 *Int'l Legal Materials* 118 (1969). For discussion see chapter 3, 50.
95 Decision of 12 February 1968, quoted in ibid, at 138, 140
96 Agreement on Settlement of Claims Relating to Gut Dam, 18 November 1968, [1968] 6 U.S.T. 7863, T.I.A.S. no. 6624
97 Communication of 27 September 1968, quoted in 8 *Int'l Legal Materials* 140–2 (1969)
98 On the problems, see Borders, Manitoba Fears River Plan in US, *N.Y. Times*, 25 August 1974, § 1, at 11, col. 1.
99 See Garrison Diversion Unit, Canada Department of External Affairs Press Release no. 48 (24 June 1975). The note cites previous diplomatic correspondence and continuing consultation going back to 1969.

100 11 January 1909, 36 Stat. 2448 (1909–11), T.S. no. 548, art. 4. See chapter 3, 43–4, 64–5.
101 IJC, Transboundary Implications of the Garrison Diversion Unit (1977). For discussion, see also Annual Report 1976 of the International Joint Commission 25–6 (1977).
102 For mention of some others see Borders, supra note 98.
103 [1949] I.C.J. 4; see chapter 3, 48, 49.
104 (Spain v France), 12 U.N.R.I.A.A. 281 (1957), digested in 53 Am. J. Int'l L. 156 (1959); see chapter 3, 48, 49–50.
105 [1949] I.C.J. 23
106 12 U.N.R.I.A.A. 303; see chapter 3, 50.
107 See McDougal and Schlei, The Hydrogen Bomb Tests in Perspective: Lawful Measures for Security, in M. McDougal and Associates, Studies in World Public Order 763 (1960).
108 Agreement on Personal and Property Damage Claims, 4 January 1955, [1955] 1 U.S.T. 1, T.I.A.S. no. 3160
109 The terms and conditions were specified in a note from the US Department of State to the Canadian government on 13 November 1974. On reimbursement for the clean-ups, see ARCO Pays $19,000 Cleanup Bill and Oil Spill Billing Will Hit $26,000, supra note 36.
110 I. Brownlie, supra note 2, at 463. See also Jenks, Liability for Ultra-Hazardous Activities, 117 Recueil des cours 166 (1966); Kelson, State Responsibility and the Abnormally Dangerous Activity, 13 Harv. Int'l L.J. 197 (1972).
111 Brownlie, supra note 110, at 463
112 25 May 1962, in 57 Am. J. Int'l L. 268, art. 2 (1963). See chapter 3, 33.
113 Brussels Nuclear Ships Liability Convention, supra note 112, art. 2
114 Ibid, art. 3. The limit was set at 1500 million francs.
115 Opened for signature 21 May 1963, in 2 Int'l Legal Materials 727, art. 4 (1963)
116 Ibid
117 Ibid
118 On this subject see generally Cigoj, International Regulation of Civil Liability for Nuclear Risk, 14 Int'l & Comp. L.Q. 809 (1965); Hardy, The Liability of Operators of Nuclear Ships, ibid, vol. 12, at 778 (1963) and his Nuclear Liability: The General Principles of Law and Further Proposals, 36 Brit. Y.B. Int'l L. 223 (1960); Konz, The 1962 Brussels Convention on the Liability of Operators of Nuclear Ships, 57 Am. J. Int'l L. 100 (1963).
119 Done 29 July 1960, in 8 Europ. Y.B. 202 (1960)
120 Convention Supplementary to the (OEEC) Paris Convention of 1960, 30 January 1963, in 2 Int'l Legal Materials 685 (1963)

121 *Done* 17 December 1971, in ibid, vol. 11, at 277 (1972)
122 Treaty on Principles Governing the Activities of States in the Exploration and Use of Outer Space, Including the Moon and Other Celestial Bodies, *done* 27 January 1967, [1967] 3 U.S.T. 2410, T.I.A.S. no. 6347, 610 U.N.T.S. 205, art. 7. See chapter 3, 39.
123 *Done* 29 March 1972, [1973] 2 U.S.T. 2389, T.I.A.S. no. 7762, art. 2. See chapter 3, 39–40.
124 Liability Convention, supra note 123, art. 4
125 Ibid, art. 3
126 Supra note 8, para. 7. See 145 supra.
127 *Done* 29 November 1969, in 9 *Int'l Legal Materials* 45 (1969). See also Protocol to International Convention on Civil Liability for Oil Pollution Damage, *done* 19 November 1976, in 16 *Int'l Legal Materials* 617 (1977); the protocol converts the limit into special drawing rights (SDRs) as defined by the International Monetary Fund as the unit of account. The 'Private Law' Convention was discussed chapter 3, 34.
128 See Healy, The CMI and IMCO Draft Conventions on Civil Liability for Oil Pollution, 1 *J. Maritime L. & Com.* 93, 93–8 (1969); Goldie, supra note 89, at 314–17. See also Avins, Absolute Liability for Oil Spillage, 38 *Brooklyn L. Rev.* 359 (1970); Bergman, No Fault Liability for Oil Pollution Damage, 5 *J. Maritime L. & Com.* 1 (1973). See generally Dowd, Further Comment on the Civil Liability and Compensation Fund Conventions, ibid, vol. 4, at 525 (1973).
129 Supra note 127, art. 3
130 In 9 *Int'l Legal Materials* 66, 67 (1970)
131 *Done* 18 December 1971, ibid, vol. 11, at 284 (1972). For analysis of the working arrangements involved, see Hunter, The Proposed International Compensation Fund for Oil Pollution Damage, 4 *J. Maritime L. & Com.* 117 (1972).
132 The limit was set at 450 million francs. On the problem of conversion rates, see Mendelsohn, Value of the Poincaré Gold Franc in Limitation of Liability Conventions, 5 *J. Maritime L. & Com.* 125 (1973). See also Protocol to the International Convention on the Establishment of an International Fund for Compensation for Oil Pollution Damage, *done* 19 November 1976, in 16 *Int'l Legal Materials* 621 (1977), which converts the limit into SDRs, supra note 127, as the unit of account.
133 See 1971 Fund Convention, supra note 131, art. 3; see also chapter 3, 34.
134 Fund Convention, supra note 31, art. 10
135 Ibid, art. 7, incorporating by reference art. 9 of the Liability Convention, supra note 127

136 7 January 1969, in 8 *Int'l Legal Materials* 497 (1969). The owners of over 99 per cent of the free world's tanker tonnage are parties (booklet entitled *TOVALOP* 4 (Int'l Tanker Owners Pollution Federation Ltd, reprint 1973)).

137 14 January 1971, in 10 *Int'l Legal Materials* 137 (1971). The receivers of over 90 per cent of the world's cargoes of crude and fuel oil contracted to be parties to the agreement (Becker, A Short Cruise on the Good Ships TOVALOP and CRISTAL, 5 *J. Maritime L. & Com.* 609, 614 (1974)).

138 TOVALOP and CRISTAL became effective on 6 October 1969 and 1 April 1971 respectively. When the former came into operation, at least 50 per cent of the tanker tonnage of the world had become parties (*TOVALOP*, supra note 136, at 4). And the latter required that Oil Companies receiving over 50 per cent of the world's seaborne crude oil and fuel oil become signatories in order to come into effect (CRISTAL, supra note 137, clause III(A)).

139 Supra note 136, art. 4

140 Ibid, art. 6

141 Supra note 137, art. 4

142 Ibid

143 The new limits of TOVALOP are to be $147 per gross registered ton or $16.8 million. Revised CRISTAL limits will match the Fund Convention $36 million ceiling. TOVALOP will also in the future include coverage for third-party damage.

144 *Done* 4 September 1974, in 13 *Int'l Legal Materials* 1409 (1974)

145 Ibid, Clause IV

146 Ibid, Clause II. The Rules of the Association appear in 14 *Int'l Legal Materials* 147 (1975).

147 *Done* 17 December 1976, in 15 *Int'l Legal Materials* 1451 (1977). This convention, which covers both public and private damage (including preventive measures), sets a limit of liability at 22 million SDRs, supra note 127. For discussion of its purposes and operation, see Dubais, The 1976 London Convention on Civil Liability for Oil Pollution Damage from Offshore Installations, 9 *J. Maritime L. & Com.* 61 (1977). For an excellent review of the whole developing law in the area of prevention of pollution from activities concerned with the exploration and exploitation of the continental shelf, see de Mestral, Study of Offshore Mining and Drilling Carried Out within the Limits of National Jurisdiction, U.N. Doc. UNEP/WG.14/2 (23 February 1978).

148 43 U.S.C.A. § 1651 (Supp. 1975)

149 Ibid, § 1653(a) and (b)

150 Ibid, § 1653(c)(1)

151 Ibid, § 1653(c)(2) and (3)

152 Ibid, § 1653(c)(5)

153 See also § 1654 of the Act, which authorizes the President of the United States to enter into negotiations with the government of Canada on a whole range of issues concerned with pipelines or other transportation systems for the transport of natural oil and gas, including environmental and energy issues.

154 495 F.2d 213 (6th Cir. 1974), *cert. denied* 419 U.S. 997, 95 S. Ct. 310, 42 L. Ed. 2d 270 (1974). Thirty-seven residents of Canada brought this suit against the three corporations, claiming that pollutants emitted by the defendants' plants were noxious and represented a nuisance which resulted in damage to their persons and property. For discussion of why this form of action was preferred to the international claims route, see Ianni, International and Private Actions in Transboundary Pollution, 11 *Can. Y.B. Int'l L.* 258, 266–70 (1973). See also chapter 7, 194. See generally McCaffrey, Transboundary Pollution Injuries: Jurisdictional Considerations in Private Litigation between Canada and the United States, 3 *Cal. Western Int'l L.J.* 191 (1973).

155 See 164–6 supra.

156 Convention on the Protection of the Environment, *done* 19 February 1974, in 13 *Int'l Legal Materials* 591 (1974). For discussion see chapter 3, 53, chapter 4, 91.

157 It was announced on 16 January 1967 by US Embassy sources in Madrid that the United States had paid $558,104 to 475 Spaniards who suffered damage when the four bombs fell. US officials said that 597 claims had been filed and that all of them would be paid. A group of residents from Palomares alleged, however, that only 3 per cent of the claims had been paid in full, and they asserted that outstanding claims totaled $2.5 million (US Pays Spanish Claims for Damage by Lost Bomb, *N.Y. Times*, 16 January 1967, at 14, col. 3).

158 See ARCO Sets Up Offices for Oil Damage Bills, supra note 36.

159 For discussion of the *Torrey Canyon* litigation see Brown and Comment, supra note 34.

CHAPTER SEVEN

1 Informal Composite Negotiating Text (ICNT), Part XV, arts. 286–97, and Annexes IV–VII, in 8 *Third United Nations Conference on the Law of the Sea: Official Records* 45–8, 57–63 (1977). There are also certain other provisions in the text relating to this subject, in particular art. 59 on the bases for resolution of conflicts regarding the attribution of rights and jurisdiction in the exclusive economic zone and Part XI § 6 (arts. 187–92) concerning the settlement of disputes relating to the international Area (ibid, at 13, 34).

2 (R. Stein ed forthcoming). For a comprehensive review of this subject, see A. Levin, The Protection of the Human Environment: Procedures and Principles for Preventing and Resolving International Controversies (UNITAR 1978).

3 See chapter 6, 'Environment monitoring' (149–50) and 'Warning and notification' (159–62).

4 For a description of Earthwatch, with its Global Environmental Monitoring System (GEMS) and the International Referral System for Sources of Environmental Information (IRS), see chapter 4, 76–8.

5 Article 34 states the limitation: 'Only States may be parties in cases before the Court.' In the *Reparations* case, the Court held that this competence extended to international organizations created by states ([1949] I.C.J. 174).

6 The Hague Peace Conferences of 1899 and 1907 each produced such a convention: 29 July 1899, 32 Stat. 1799 (1901–3), T.S. no. 392; 18 October 1907, 36 Stat. 2199 (1909–11), T.S. no. 536, respectively.

7 26 September 1928, 93 L.N.T.S. 345

8 European Convention for the Peaceful Settlement of Disputes, *done* 29 April 1957, 320 U.N.T.S. 243; American Treaty on Pacific Settlement ('Pact of Bogotá'), 30 April 1948, 30 U.N.T.S. 55

9 For examples see UN, *Systematic Survey of Treaties for the Pacific Settlement of International Disputes, 1928–1948* (1948). See also M. Habicht, *Post-War Treaties for the Pacific Settlement of International Disputes* (1931).

10 (*United States* v *Canada*), 3 U.N.R.I.A.A. 1911 and 1938 (1938 and 1941); (*United States* v *Canada*), reported in Canada-United States Settlement of Gut Dam Claims, 22 September 1968, Report of the Agent of the United States before the Lake Ontario Claims Tribunal, in 8 *Int'l Legal Materials* 118 (1969). For discussion of these two arbitrations see chapter 3, 48–9 and 50, chapter 6, 164–6.

11 (*Australia* v *France*) and (*New Zealand* v *France*), [1974] I.C.J. 253 and 457. For discussion see chapter 3, 41–2, chapter 4, 91–2, chapter 6, 147, 158.

12 See chapter 3, 47. Presentation of facts and citation of references in both these spills is found at chapter 6, 151–2 and 174–5.

13 The increase in amounts of anthropogenic sulphur in the air over Western Europe due to industrial discharges has allegedly led to greater acidity of the rains and caused consequent damage to a number of lakes in Norway and Sweden. See Air Pollution Across National Boundaries: The Impact on the Environment of Sulphur in Air and Precipitation (Swedish case study for the UNCHE 1971); Bolin's piece of the same title in *International Responsibility for Environmental Injury* (R. Stein ed forthcoming).

14 For details see chapter 6, 155–6.

15 42 U.S.C. § 1857f-2(b)(2) (1970). And, of course, European and other countries have developed their own national or regional standards, further complicating transnational sales programs.

16 See IBRD, *Environmental, Health, and Human Ecologic Considerations in Economic Development Projects* (1973); AID, *Feasibility Studies, Economic and Technical Soundness Analyses, Capital Projects* (1964, addendum 1972). See also AID, Environmental Procedures, 22 C.F.R. Part 216. See generally IIED, *Multilateral Aid and the Environment* (1977).
17 For definition and discussion see M. McDougal, H. Lasswell, and I. Vlasic, *Law and Public Order in Space* 735 et seq. (1963). Most writers use the labels 'direct' and 'indirect,' but these authors prefer the terms 'primary' and 'secondary' in view of the fact that not only judicial decisions but also executive and legislative acts may require subsequent efforts for their implementation.
18 Article 33 reads: '1 / The parties to any dispute, the continuance of which is likely to endanger the maintenance of international peace and security, shall, first of all, seek a solution by negotiation, enquiry, mediation, conciliation, arbitration, judicial settlement, resort to regional agencies or arrangements, or other peaceful means of their own choice. 2 / The Security Council shall, when it deems necessary, call upon the parties to settle their dispute by such means.' These means may be, of course, applicable to the settlement of all types of international disputes – not only those immediately threatening 'peace and security.' For the pervasive purposes of international environmental law, it also seems unnecessary to distinguish between an actual 'dispute' and a 'situation which might lead to international friction or give rise to a dispute,' as is done in the context of chapter 6 of the Charter.
19 See chapter 3, 48, chapter 6, 164, *et seq.*
20 See W. Bishop, *International Law* 842–51 (1971); 6 G. Hackworth, *International Law* 142–9 (1961); R. Lillich, *International Claims: Their Adjudication by National Commissions* (1962).
21 For discussion see F. Iklé, *How Nations Negotiate* (1964); A. Lall, *Modern International Negotiation* (1966).
22 J. Stone, *Legal Controls of International Conflict* 67 (2d ed rev 1959)
23 Agreement Approving Minute 242 of the International Boundary and Water Commission Setting Forth a Permanent and Definitive Solution to the International Problem of the Salinity of the Colorado River, *done* and *entered into force* 30 August 1972, [1973] 2 U.S.T. 1968, T.I.A.S. no. 7780. As regards the salinity problem and the course of its resolution see chapter 3, 44.
24 For discussion of the Boundary Waters Treaty of 1909, the International Joint Commission, and related matters see chapter 3, 43–4. See also L. Bloomfield and G. Fitzgerald, *Boundary Waters Problems of Canada and the United States* (1958).
25 See Polish-Czechoslovak Bilateral Cooperation in the Area of Water Pollution Control (Polish case study for the UNCHE 1971).

26 For a note on the Rhine and Danube Commissions and their limitations in dealing with environmental problems, see chapter 3, 223 n146; see also Bourne, The Waters of International Drainage Basins, in *Avoidance and Adjustment of Environmental Disputes*, supra note 2. On the role of fisheries commissions, the ICNAF example is discussed at chapter 3, 57; see generally D. Johnston, *The International Law of Fisheries* (1965); A. Koers, *International Regulation of Marine Fisheries* (1973).

27 See OECD, Council Recommendation on Principles Governing Transfrontier Pollution, in 14 *Int'l Legal Materials* 242 (1975). See also OECD, Council Recommendation on Implementing a Regime of Equal Right of Access and Non-Discrimination in Relation to Transfrontier Pollution, ibid, vol. 16, at 977 (1977). See generally OECD, *Problems in Transfrontier Pollution* (1974).

28 It will also be remembered that an extant dispute between Argentina and Brazil underlay the failure of the UN Human Environment Conference to reach agreement on 'draft Principle 20' dealing with exchange of environmental information. See chapter 3, 50-1, chapter 6, 159-61.

29 See, eg, the Pact of Bogotá, supra note 8, art. 9.

30 See, eg, W. Bishop, supra note 20, at 63; cf L. Goodrich and A. Simons, *The United Nations and the Maintenance of International Peace and Security* 291 (1955) and the multilateral pacific settlement treaties, supra notes 6-8. See, eg, the Pact of Bogotá, supra note 8, arts. 11-12.

31 Despite the variations in nomenclature, the underlying functions are comparable enough for present purposes. For a delineation of the steps involved, see, eg, the Hague Convention of 1907, supra note 6, arts. 9-14.

32 See note 28 supra.

33 The Geneva Convention on Fishing and Conservation of the Living Resources of the High Seas, *done* 29 April 1958, [1966] 1 U.S.T. 138, T.I.A.S. no. 5969, 559 U.N.T.S. 284, art. 9, for instance, specifies that any dispute arising under certain of its provisions, at the request of any of the parties, shall be 'submitted for settlement to a special commission of five members, unless the parties agree to seek a solution by another method of peaceful settlement, as provided for in Article 33 of the Charter of the United Nations.'

34 These matters are dealt with in ICNT art. 296, in 8 *Third UNCLOS: Official Records*, supra note 1, at 48. The compulsory conciliation formula was agreed upon at the seventh session of UNCLOS III in Negotiating Group 5. See Results of the Work of the Negotiating Group on Item (5) of Document A/Conf.62/62: Report to the Plenary by the Chairman, Ambassador Constantin Stavropoulos, in *Reports of the Committees and Negotiating Groups on Negotiations at the Seventh Session contained in a Single Document Both for the Purposes of Record and for the Convenience of Delegations* 100-6 (19 May 1978). This for-

mula was a compromise between those states who wanted a compulsory and binding resolution of all allegations of abuse of coastal state rights and those who did not want their sovereign rights subject to any such procedures. Compulsory conciliation, while something of a novel concept, has a venerable precedent in the Vienna Convention on the Law of Treaties *done* 23 May 1969, in 8 *Int'l Legal Materials* 679, art. 66 and the Annex (1969). For discussion of the interrelationship of this issue with other jurisdictional matters, see chapter 3, 213 n35.

35 See Resolution on Institutional and Financial Arrangements, in *Report of the United Nations Conference on the Human Environment*, U.N. Doc. A/Conf.48/14, at 61, 63 (1972).

36 ICNT art. 284, in 8 *Third UNCLOS: Official Records*, supra note 1, at 46, deals with conciliation. The compulsory conciliary formula, which is art. 296 bis, is discussed supra note 34. Annex IV of the ICNT, ibid, at 57, also deals with conciliation. Compare Annexes V and VI, ibid, at 58 and 61, on the Statute of the Law of the Sea Tribunal and on Arbitration. See also Annex VII, ibid, at 62, which provides for special arbitration procedures for, *inter alia*, disputes relating to 'protection and preservation of the marine environment' (Art. 1, ibid).

37 For a comprehensive analysis of the precedents in this regard, see Schwebel and Welter, Arbitration and the Exhaustion of Local Remedies, 60 *Am. J. Int'l L.* 484 (1966); and for discussion specifically of the non-exhaustion of anterior processes, see W. Reisman, *Nullity and Revision* 359-75 (1971).

38 *Lex loci contractus* is a doctrine establishing a preference for the law of the place where an agreement is made, but is often mitigated by reference to the law of the places of performance, domicile, nationality, or unexpressed preference of the parties. Yet by far the most international environmental claims tend to be of a tortious genre, and the usual solution is that of *lex loci delicti commissi* (interpreted as the law of the place where the act[s] giving rise to the liability first occasioned injurious effect). Many acts producing environmental injury, however, are likely to occur in common areas, originating aboard vessels or spacecraft. Then factors such as the law of the flag or state of registration, the allegiance or domicile of the injured party, the allegiance of the defendant craftowner, and so forth can come into consideration. The situation can be further complicated where resort is to national rather than international tribunals. The state of the forum may reject the application of *lex loci delicti* and other prescriptive competences in favour of its own law by regarding the question as merely a matter of 'procedure,' thereby circumventing substantive conditions, or alternatively as an important matter of 'public policy,' abnegating the requirement of deference to the acts of state of another. The

classic case discussing these problems is *Lauritzen* v *Larsen*, 345 U.S. 571, 73 S. Ct. 921, 97 L.Ed. 1254 (1953). See generally *Restatement (Second) Conflict of Laws* (1969); W. Reese and M. Rosenberg, *Cases and Materials on Conflict of Laws* (1971).

39 Supra note 6, art. 37
40 On the compromisory process, see W. Reisman, supra note 37, at 75–106.
41 Supra note 6
42 Rules of Arbitration and Conciliation for Settlement of International Disputes Between Two Parties of Which Only One Is a State, in 57 *Am. J. Int'l L*. 500 (1963)
43 See, eg, Convention on International Trade in Endangered Species of Wild Flora and Fauna, *done* 3 March 1973, [1976] 2 U.S.T. 1087, T.I.A.S. no, 8247; International Convention for the Prevention of Pollution from Ships, *done* 2 November 1973, in 12 *Int'l Legal Materials* 1319, art. 10 (1973).
44 Supra note 10
45 The IJC Report on the Trail Smelter Investigation, Docket no. 25, was unanimous and was rendered on 28 February 1931.
46 T.S. no. 893, 49 Stat. 3245 (1935–6), reprinted in 3 U.N.R.I.A.A. 107–10
47 3 U.N.R.I.A.A. 1911 (1938), reprinted in 33 *Am. J. Int'l L*. 182 (1939)
48 3 U.N.R.I.A.A. 1938 (1941), reprinted in ibid, vol. 35, at 684 (1941)
49 3 U.N.R.I.A.A. 1980
50 3 U.N.R.I.A.A. 1965. See chapter 3, 48–9.
51 [1949] I.C.J. 4. For previous discussion of the substance see chapter 3, 49, chapter 6, 147, 166–7.
52 Reprinted in the decision, at 12 U.N.R.I.A.A. 285–6
53 12 U.N.R.I.A.A. 303. See chapter 3, 49–50.
54 Supra note 10
55 Act of 15 August 1962, Pub. L. no. 87–587, 76 Stat. 387
56 Agreement Concerning the Establishment of an International Arbitral Tribunal to Dispose of U.S. Claims Relating to Gut Dam, 25 March 1965, [1966] 2 U.S.T. 1566, T.I.A.S. no. 6114, reprinted in 4 *Int'l Legal Materials* 468 (1965)
57 See The Foreign Claims Settlement and the Lake Ontario Claims Program, in 4 *Int'l Legal Materials* 473 (1965).
58 See ibid, vol. 8, at 136 (1969).
59 Ibid, at 138–40
60 Ibid, at 140
61 Agreement on Settlement of Claims Relating to Gut Dam, 18 November 1968, [1968] 6 U.S.T. 7863, T.I.A.S. no. 6624
62 8 *Int'l Legal Materials* 141–2
63 Ibid, at 118

64 ICNT Annex VI, in 8 *Third UNCLOS: Official Records*, supra note 1, at 61
65 See note 28 and 183 supra.
66 Article 38 reads in full: '1 / The Court, whose function is to decide in accordance
with international law such disputes as are submitted to it, shall apply: a / international conventions, whether general or particular, establishing rules expressly
recognized by the contesting states; b / international custom, as evidence of a
general practice accepted as law; c / the general principles of law recognized by
civilized nations; d / subject to the provisions of Article 59 [no binding force except between the parties and in respect of that particular case], judicial decisions
and the teachings of the most highly qualified publicists of the various nations, as
subsidiary means for the determination of rules of law. 2 / This provision shall not
prejudice the power of the Court to decide a case *ex aequo et bono* if the parties
agree thereto.'
67 Jessup, Do New Problems Need New Courts? 65 *Proc. Am. Soc. Int'l L.* 261
(1971)
68 Ibid, at 262-3. Although the use of chambers is expressly provided for in arts. 26
to 29 of the ICJ Statute, they have never yet been employed. See also art. 50 on
procedures for carrying out inquiries and obtaining expert opinions.
69 Jessup, supra note 67, at 263-4
70 Ibid, at 265. See also 178-9 and note 5 supra.
71 See chapter 6, 157-8.
72 [1949] I.C.J. 4. For previous discussion of the substance see chapter 3, 49, chapter 6,
147, 166-7.
73 The Security Council resolution of 9 April 1947 is quoted in *Corfu Channel* case,
Judgment on Preliminary Objection, [1948] I.C.J. 15, 17. The Council dealt with
this matter under art. 36 of the U.N. Charter. See also ibid, art. 33, quoted in
note 18 supra.
74 The proceedings were instituted on 22 May 1947. See Application Instituting
Proceedings, 1 *Corfu Channel* case, I.C.J. Pleadings 8 (1947).
75 Judgment on Preliminary Objection, [1948] I.C.J. 15
76 Order of 26 March 1948, [1948] I.C.J. 53
77 [1949] I.C.J. 4
78 Judgment of 15 December 1949, [1949] I.C.J. 244
79 (*United Kingdom* v *Iceland*) and (*Federal Republic of Germany* v *Iceland*),
[1974] I.C.J. 3 and 175. For prior substantive discussion see chapter 3, 58-9,
chapter 6, 157-8.
80 (*Australia* v *France*) and (*New Zealand* v *France*), [1974] I.C.J. 253 and 457.
For prior discussion see chapter 3, 41-2, chapter 4, 91-2, chapter 6, 147, 158.
81 [1973] I.C.J. 3 and 49
82 [1974] I.C.J. 3

83 See chapter 6, 157–8. The Iceland–Great Britain Interim Agreement in the Fisheries Dispute, *done* 13 November 1973, appears in 12 *Int'l Legal Materials* 1315 (1973).

84 [1974] I.C.J. 253 and 457

85 Order of 20 December 1974, [1974] I.C.J. 530. See chapter 6, 257 n22.

86 See also 199 infra.

87 For discussion see chapter 6, 158.

88 [1974] I.C.J. 253 and 457

89 For a description of the process and relevant provisions, see H. Steiner and D. Vagts, *Transnational Legal Problems* 1065–6 (1968). See also Jackson, GATT as an Instrument for the Settlement of Trade Disputes, 61 *Proc. Am. Soc. Int'l L.* 144 (1967).

90 *Opened for signature* 27 August 1965, [1966] 1 U.S.T. 1270, T.I.A.S. no. 6090, 575 U.N.T.S. 159. For commentary see Broches, The Convention on the Settlement of Investment Disputes: Some Observations on Jurisdiction, 5 *Colum. J. Transnat'l L.* 263 (1966); Rodley, Some Aspects of the World Bank Convention on the Settlement of Investment Disputes, 4 *Can. Y.B. Int'l L.* 43 (1966). See also H. Steiner and D. Vagts, supra note 89, at 360–3.

91 See 181 supra.

92 The European Convention for the Protection of Human Rights and Fundamental Freedoms, 4 November 1950, Eur. T.S. no. 5, 213 U.N.T.S. 221, created two organs: the European Commission for Human Rights and the Court of the same. Any individual may file an application with the Commission alleging deprivation of a right guaranteed under the Convention, and the Commission or an interested state party may take the case to the Court under specified conditions. The European Court of Human Rights has itself had a rather unimpressive caseload, but the Convention and the possibility of such adjudication can be seen to have had a not inconsiderable influence on the actions of national judiciaries (H. Steiner and D. Vagts, supra note 89, at 302–3). See generally Golsong, The Control Machinery of the European Convention on Human Rights, *Int'l & Comp. L.Q.* Supp. no. 11, at 38 (1965) and his The European Convention on Human Rights Before Domestic Courts, 38 *Brit. Y.B. Int'l L.* 445 (1962).

93 See 2 *United Nations Conference on the Law of the Sea: Official Records* 145–6, U.N. Doc. A/Conf.13/38 (1958).

94 See the alternatives presented in 5 *Report of the Committee on the Peaceful Uses of the Sea-Bed and the Ocean Floor Beyond the Limits of National Jurisdiction*, 28 U.N. GAOR Supp. 21, appendix § 21, at 3, U.N. Doc. A/9021 (1973).

95 See 177 and 183 supra.

96 ICNT Annex V, in 8 *Third UNCLOS: Official Records*, supra note 1, at 58–60. The new Tribunal will feature a special Sea-Bed Disputes Chamber composed

of eleven of its members, also selected to represent principal legal systems and equitable geographical distribution (art. 37, ibid, at 60).

In general, a state upon becoming party to the new convention will have to choose and declare one or more means of dispute settlement: the new Law of the Sea Tribunal, the International Court of Justice, an arbitral tribunal constituted in accordance with the terms of the convention, or a special arbitral tribunal (ICNT art. 287, ibid, at 46). There are, however, special provisions as to jurisdiction of the Sea-Bed Disputes Chamber for disputes between the new International Authority and other parties (ICNT arts. 187–92, ibid, at 34). See also 177 supra.

97 ICNT Annex V, art. 2(1) and (2), in 8 *Third UNCLOS: Official Records*, supra note 1, at 58

98 Annex V, art. 23, ibid, at 59

99 As, for example, Mr Justice Gray explained in the Paquete Habana: 'International law is part of our law, and must be ascertained and administered by the courts of justice of appropriate jurisdiction, as often as questions of right depending upon it are duly presented for their determination. For this purpose, where there is no treaty, and no controlling executive or legislative act or judicial decision, resort must be had to the customs and usages of civilized nations; and, as evidence of these, to the works of jurists and commentators' (175 U.S. 677, 700, 20 S. Ct. 290, 44 L.Ed. 320 (1900)). See generally McDougal, The Impact of International Law upon National Law: A Policy-Oriented Perspective, 4 *S. Dak. L. Rev.* 25 (1959).

100 On this problem, see W. Reisman, supra note 37, at 359–75.

101 See chapter 6, 155, 156–7.

102 404 U.S. 917, 92 S. Ct. 242, 30 L.Ed. 2d 191 (1971). For opinions below see *Committee for Nuclear Responsibility v Seaborg*, 463 F.2d 783 (D.C. Cir. 1971), 463 F.2d 788 (D.C. Cir. 1971), 463 F.2d 796 (D.C. Cir. 1971).

103 479 F.2d 842 (D.C. Cir. 1973). For an earlier stage of the litigation, see *Wilderness Soc'y v Hickel*, 325 F. Supp. 422 (D.D.C. 1970). See also *Natural Resources Defense Council v Morton*, 458 F.2d 827 (D.C. Cir. 1972).

104 *Wilderness Soc'y v Morton*, 463 F.2d 1261, 1262–3 (D.C. Cir. 1972)

105 *Michie v Great Lakes Steel Div., Nat'l Steel Corp.*, 495 F.2d 213 (6th Cir. 1974), *cert. denied* 419 U.S. 997, 95 S. Ct. 310, 42 L.Ed. 2d 270 (1974). See chapter 6, 174.

106 Ianni, International and Private Actions in Transboundary Pollution, 11 *Can. Y.B. Int'l L.* 258, 267 (1973)

107 Ibid, at 264–6

108 See ibid, at 269–70.

109 The case in fact survived two separate preliminary motions for dismissal. For discussion see 495 F.2d 215; Ianni, supra note 106, at 267.

110 *Done* 19 February 1974, in 13 *Int'l Legal Materials* 591 (1974). For discussion see chapter 3, 53, chapter 4, 91. But one writer has done a comprehensive study on the problems of bringing an environmental action in one of these countries, Norway (Fleischer, Eva Funder, *Miljoloven: Enhver har rett til ren luft, ren jord og rent vann* (Environmental Law: Everyone has the right to clean air, clean soil, and clean water), Norway 1978).

111 Nordic Convention, supra note 110, art. 2

112 Ibid, art. 3

113 Ibid, art. 4

114 Ibid

115 Ibid, arts. 11 and 12

116 See 180 supra.

117 On national courts see R. Falk, *The Role of Domestic Courts in the International Legal Order* (1964); W. Friedmann, *The Changing Structure of International Law* 141–8 (1964); cf W. Reisman, supra note 37, at 802–35.

118 A well-known example of such rigid adherence to the act of state doctrine by the US Supreme Court is, of course, *Banco Nacional de Cuba* v *Sabbatino*, 376 U.S. 398, 84 S. Ct. 923, 11 L.Ed. 2d 804 (1964).

119 See 188–9 supra.

120 See Schachter, The Enforcement of International Judicial and Arbitral Decisions, 54 *Am. J. Int'l L.* 1 (1960); see generally W. Reisman, supra note 37.

121 See W. Reisman, supra note 37, at 732–53.

122 Ibid, at 172–3

123 In connection with this tangled polemic of international environmental dispute settlement procedures, one is reminded of the summary conclusion arrived at by a prominent professor: 'Successive limited comparisons is, then, indeed a method or system; it is not a failure of method for which administrators ought to apologize ... Why the bother to describe the method in all the above detail? Because it is in fact a common method of policy formulation, and is, for complex problems, the principal reliance of administrators as well as of other policy analysts. And because it will be superior to any other decision-making method available for complex problems in many circumstances, certainly superior to a futile attempt at superhuman comprehensiveness' (Lindblom, The Science of 'Muddling Through,' 19 *Pub. Ad. Rev.* 79, 87–8 (1959)).

124 But see the elaborate provisions for compulsory resort to adjudication and conciliation under the new Vienna Convention on the Law of Treaties, supra note 34. For discussion of the dispute settlement provisions of this convention, see Kearney and Dalton, The Treaty on Treaties, 64 *Am. J. Int'l L.* 495, 545–57 (1970).

125 A choice by states among the alternatives mentioned is provided for in ICNT art. 287, in 8 *Third UNCLOS: Official Records*, supra note 1, at 46. On the compulsory conciliation provisions, see 183 supra.

126 See article 38 of the Statute of the International Court of Justice, quoted in note 66 supra.

127 The US declaration recognizing compulsory jurisdiction under article 36(2) of the ICJ Statute, in relevant part, explicitly specifies the exception: '*Provided*, that this declaration shall not apply to ... b / Disputes with regard to matters which are essentially within the domestic jurisdiction of the United States of America as determined by the United States of America' (*done* 14 August 1946, 1 U.N.T.S. 10).

128 The French reservation reads in its relevant part as follows: 'with the exception of: ... 3 / Disputes arising out of a war or international hostilities, disputes arising out of a crisis affecting national security or out of any measure or action relating thereto, and disputes concerning activities connected with national defence' (16 May 1966, 562 U.N.T.S. 71).

129 9 April 1970, in 9 *Int'l Legal Materials* 598 (1970). See also chapter 4, 85–6 *et seq.*

130 In the *Interhandel* case, Preliminary Objections, [1959] I.C.J. 6, 23, the Court explained: 'Reciprocity in the case of Declarations accepting the compulsory jurisdiction of the Court enables a Party to invoke a reservation to that acceptance which it has not expressed in its own Declaration.'

List of abbreviations

ACC   United Nations Administrative Committee on Co-ordination
AID   Agency for International Development (US)
ASIL   American Society of International Law
CCMS   Committee on the Challenges of Modern Society (NATO)
CEQ   Council on Environmental Quality (US)
CILSS   Permanent Inter-State Drought Control Committee for the Sahel
CMEA   Council for Mutual Economic Assistance
CRISTAL   Contract Regarding an Interim Supplement to Tanker Liability
   for Oil Pollution
ECE   United Nations Economic Commission for Europe
ECOSOC   United Nations Economic and Social Council
EEZ   exclusive economic zone
ELI   Environmental Law Institute
EPA   Environmental Protection Agency (US)
FAO   Food and Agricultural Organization
FCMA   Fisheries Conservation and Management Act (US)
FOE   Friends of the Earth
GA   United Nations General Assembly
GATT   General Agreement on Tariffs and Trade
GEMS   Global Environmental Monitoring System (UNEP)
IAEA   International Atomic Energy Agency
IBRD or 'World Bank'   International Bank for Reconstruction and Develop-
   ment
ICJ or 'World Court'   International Court of Justice
ICNAF   International Commission for Northwest Atlantic Fisheries
ICNT   Informal Composite Negotiating Text (UNCLOS)
ICSU   International Council of Scientific Unions

IIEA   International Institute for Environmental Affairs
IIED   International Institute for Environment and Development
ILA   International Law Association
ILC   International Law Commission
IJC   International Joint Commission
IMCO   International Maritime Consultative Organization
IMF   International Monetary Fund
INTELSAT   International Telecommunications Satellite Organization
IOC   International Oceanographic Commission (UNESCO, FAO, WMO, IMCO, and IAEA)
IRPTC   International Register of Potentially Toxic Chemicals
IRS   International Referral System (UNEP)
IUCN   International Union for the Conservation of Nature and Natural Resources
IWC   International Whaling Commission
LOS   *see* UNCLOS
MARPOL   Pollution from Ships Convention
MEPC   Marine Environment Protection Committee (IMCO)
NATO   North Atlantic Treaty Organization
NEPA   National Environmental Protection Act (US)
NGO   non-governmental organization
NRDC   National Resources Defense Council
OAU   Organization of African Unity
OECD   Organization for Economic Co-operation and Development
OPOL   Offshore Pollution Liability Agreement
PCIA   Permanent Court of International Arbitration
PCIJ   Permanent Court of International Justice
RFF   Resources for the Future
RIOS   River Inputs into Ocean Systems (UNEP)
RSNT   Revised Single Negotiating Text (UNCLOS)
SCEP   Study of Critical Environmental Problems (ICSU)
SCOPE   Scientific Committee on Problems of the Environment (ICSU)
SMIC   Study of Man's Impact on Climate (ICSU)
SNT   Single Negotiating Text (UNCLOS)
SOLAS   Safety of Life at Sea Convention
TOVALOP   Tanker Owners Voluntary Agreement on Liability for Oil Pollution
UN   United Nations
UNCHE or 'Stockholm Conference'   United Nations Conference on the Human Environment

UNCLOS   United Nations Conference on the Law of the Sea
UNDP   United Nations Development Programme
UNEP   United Nations Environment Programme
UNESCO   United Nations Educational, Scientific and Cultural Organization
UNIDO   United Nations Industrial Development Organization
WMO   World Meteorological Organization

Table of cases

Index of treaties and other international agreements

Name index

Subject index

abatement:
- customary international rights of 46–7
- necessary and proportional force 46–7
- and minimization of damage 150–4
absolute liability 33, 39–40, 145, 164, 168–71
abuse of rights and powers 132
access to resources 66–74, 121–4
acidic rain 179
actors 7–9, 130–1, 175, 178–9
Addis Ababa Declaration 213 n33
additionality principle 97–8, 101–02
adjudication 177, 180, 187–94, 197, 198–9
- *see also* case by name, International Court of Justice, national courts and administrative tribunals
admiralty, law of 61
aerosol sprays 117
Agency for International Development (AID) (US) 104, 179
air pollution: *see* Clean Air Act (US), *Nuclear Tests* cases, *Trail Smelter* arbitration
airspace 21, 146
- *see also* atmosphere and outer space
air transport 61–2
Albania 49, 166–7, 188–9, 257 n17

Amchitka blast: *see* Cannikin
American Society of International Law (ASIL) 177
Amoco Cadiz 152
anadromous species 27
anarchy, theories of 135
Antarctica 21, 42–3, 58, 112, 117
application of law in concrete circumstances 76, 92–6
- *see also* enforcement
appraisal of environmental decision-making, watchdog function 76, 99–101, 133–4
arbitration and arbitral tribunals 48–50, 89–90, 132, 177, 180, 184–7, 191, 197, 198, 199, 268 n10
- *see also Flegenheimer* arbitration, *Gut Dam* arbitration, *Lac Lanoux* arbitration, *Trail Smelter* arbitration
Arbitration and Conciliation for Settlement of International Disputes between Two Parties Only One of Which Is a State, Rules of 185, 243 n68
archipelagic principle 23
Arctic 27

www.ingramcontent.com/pod-product-compliance
Lightning Source LLC
Chambersburg PA
CBHW030452210326
41597CB00013B/639